Hello, happy to see you here.

I'm BB, aka Bead Baby.

Abacus is fun, if you know All About it.

"How to Abacus" has books to help you.

Book 1: All About Abacus

Book 2: All About Formulas

Book 3: All About Multiplication

Book 4: All About Division

I will definitely be with you through all the lessons.

Let's go!

To those interested in Abacus

~ Sheena

A useful tool in math learning

~ Yonhao

For information address contact@pingrow.com

ISBN: 978-1-949622-05-8

Visit www.pingrow.com

How to Abacus 4

All About Division

Division and Abacus

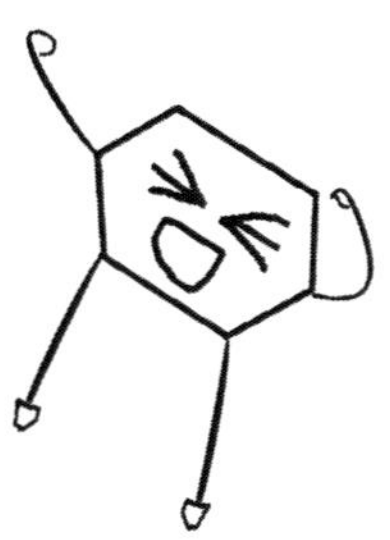

Division and Abacus

Division problems contain a serial multiplying and subtracting processes. It is necessary to establish the fundamental knowledge of multiplication, and subtraction when solving division questions with an abacus. To solve division problems correctly, you can review the other books in the series: How to Abacus 1, How to Abacus 2 and How to Abacus 3.

1. Formulas

Numbers need to be subtracted during the division process and formulas are the key to solve any subtractions on the abacus. Therefore, it is crucial to be very familiar with the formulas of Little Friends, Big Friends, and Mixed Friends that are introduced in "How to Abacus 2".

Big Friends Formulas

+ 1 = - 9 + 10	- 1 = - 10 + 9
+ 2 = - 8 + 10	- 2 = - 10 + 8
+ 3 = - 7 + 10	- 3 = - 10 + 7
+ 4 = - 6 + 10	- 4 = - 10 + 6
+ 5 = - 5 + 10	- 5 = - 10 + 5
+ 6 = - 4 + 10	- 6 = - 10 + 4
+ 7 = - 3 + 10	- 7 = - 10 + 3
+ 8 = - 2 + 10	- 8 = - 10 + 2
+ 9 = - 1 + 10	- 9 = - 10 + 1

Little Friends Formulas

+ 1 = + 5 – 4	- 1 = - 5 + 4
+ 2 = + 5 – 3	- 2 = - 5 + 3
+ 3 = + 5 – 2	- 3 = - 5 + 2
+ 4 = + 5 – 1	- 4 = - 5 + 1

Mixed Friends Formulas

+ 6 = - 5 + 1 + 10	- 6 = - 10 + 5 -1
+ 7 = - 5 + 2 + 10	- 7 = - 10 + 5 - 2
+ 8 = - 5 + 3 + 10	- 8 = - 10 + 5 - 3
+ 9 = - 5 + 4 + 10	- 9 = - 10 + 5 - 4

2. Times Table

When solving division problems, you need to use multiplication. You have to find a product that is closest to the dividend and then use the dividend to subtract from the closet product. You many repeat the steps several times to finish the division operations.

The table on next page showed the division question: 300 is divided by 4. 4 is bigger than 3 so the first step is that you need to find the closet product to 30. Based on the Multiplication chart, the closet product to 30 for 4 is 28 (4 X 7). 30 minus 28 is 2. Then you need to repeat the same process to find the closest product to 20 for 4, which is 20 (4 X 5). There is no remainder for 300 divided by 4 and the quotient is 75.

Division Question

$$300 \div 4 = 75$$

```
     75
  ______
4 ) 300
   -28
   ____
     20
     20
   ____
      0
```

You will be able to solve division problems quickly and accurately if you're familiar with the Times Table. The Times Table is on the next page so let's review and memorize the multiplication table together.

Ready, Set, Go!!!

Times Table

2 x 1 = 02	3 x 1 = 03	4 x 1 = 04	5 x 1 = 05
2 x 2 = 04	3 x 2 = 06	4 x 2 = 08	5 x 2 = 10
2 x 3 = 06	3 x 3 = 09	4 x 3 = 12	5 x 3 = 15
2 x 4 = 08	3 x 4 = 12	4 x 4 = 16	5 x 4 = 20
2 x 5 = 10	3 x 5 = 15	4 x 5 = 20	5 x 5 = 25
2 x 6 = 12	3 x 6 = 18	4 x 6 = 24	5 x 6 = 30
2 x 7 = 14	3 x 7 = 21	4 x 7 = 28	5 x 7 = 35
2 x 8 = 16	3 x 8 = 24	4 x 8 = 32	5 x 8 = 40
2 x 9 = 18	3 x 9 = 27	4 x 9 = 36	5 x 9 = 45

6 x 1 = 06	7 x 1 = 07	8 x 1 = 08	9 x 1 = 09
6 x 2 = 12	7 x 2 = 14	8 x 2 = 16	9 x 2 = 18
6 x 3 = 18	7 x 3 = 21	8 x 3 = 24	9 x 3 = 27
6 x 4 = 24	7 x 4 = 28	8 x 4 = 32	9 x 4 = 36
6 x 5 = 30	7 x 5 = 35	8 x 5 = 40	9 x 5 = 45
6 x 6 = 36	7 x 6 = 42	8 x 6 = 48	9 x 6 = 54
6 x 7 = 42	7 x 7 = 49	8 x 7 = 56	9 x 7 = 63
6 x 8 = 48	7 x 8 = 56	8 x 8 = 64	9 x 8 = 72
6 x 9 = 54	7 x 9 = 63	8 x 9 = 72	9 x 9 = 81

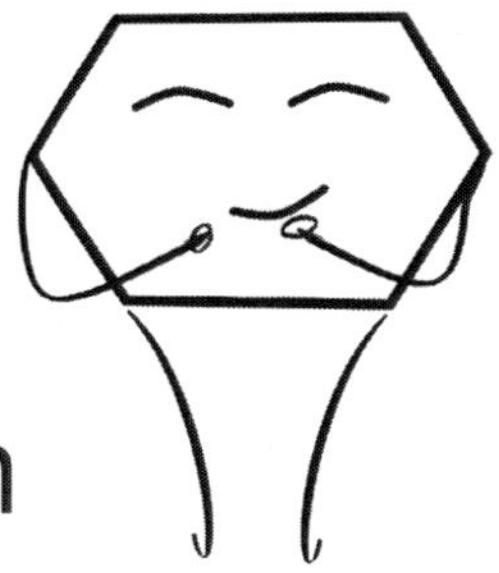

3. Leave a space for zero

Multiplication arithmetic plays an important role when solving the division problems (Please review the rules as depicting in How to Abacus 3 – All About Multiplications). Every number is important, including zero. In order to get the accurate quotient, you need to reserve a space for the number zero. If the product is one digit during the multiplication operation, you should leave a space for the number zero ex. 02, 03, 04 and so on.

The reason to leave a space for the zero number during the operations is that abacus arithmetic does not start from the ones place. Abacus algorithm always starts from the left to the right. The answers are shown on the abacus from the left to right when each step is finished.

4. Dividend on the abacus

Unlike multiplication that only puts numbers of the product on the abacus, the dividend is on the abacus before starting the division operation. But where do you put the numbers of the dividend?

The picture below shows the numbers for ones place (1), tens place (2), tenth place (0) and hundredth place (-1) on the abacus. These numbers are the clues for where to put the dividend on the abacus.

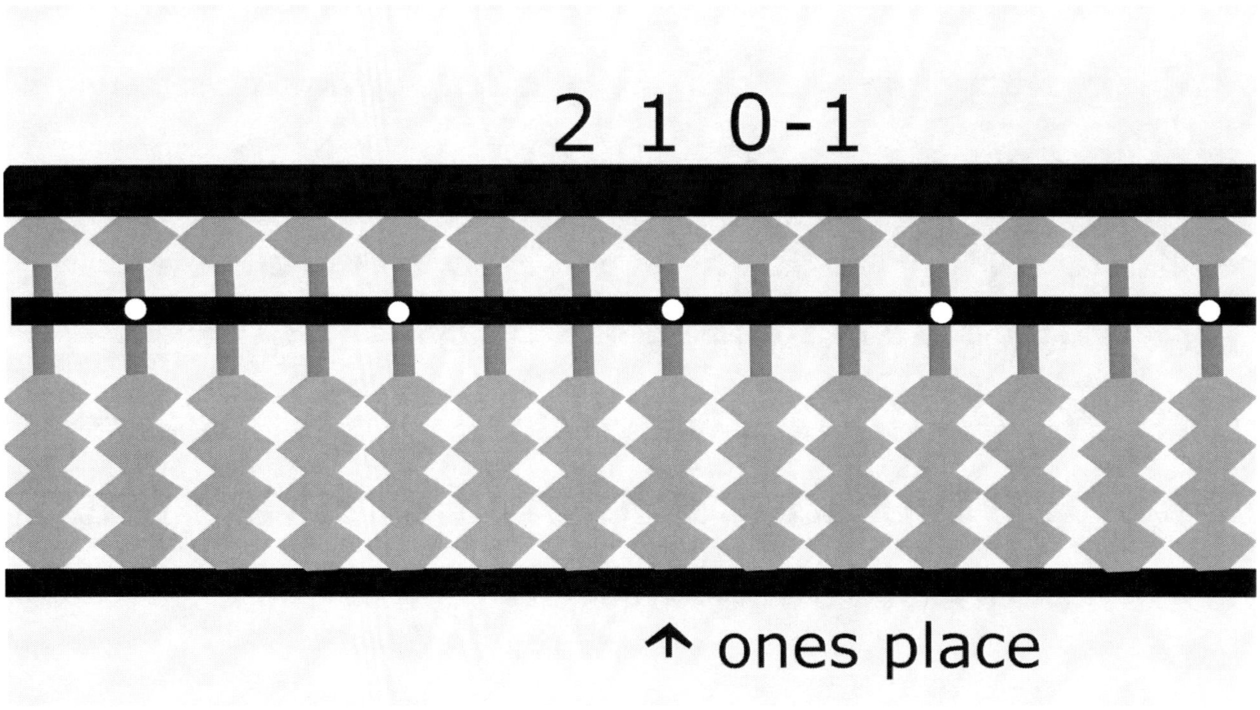

First, we need to find out how many digits the dividend and devisor have respectively. Unlike multiplication, we have to consider the division symbol as a digit. In order to find the place to put the dividend, we use the total digits of the dividend to minus the division symbol and the total digits of the devisor. (# of dividend digits - division symbol - # devisor digits). As the picture below shows, 3 digits of the dividend subtracts the division symbol and 1 digit of divisor. Therefore, the dividend should be placed in the one's place.

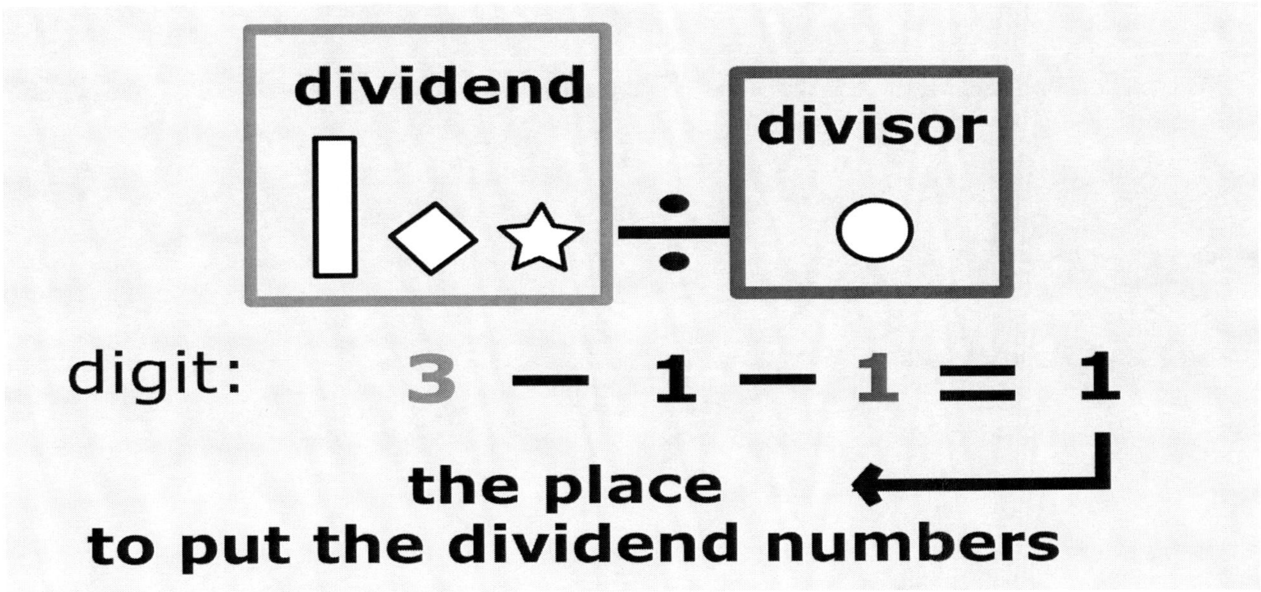

For the division question: 344 ÷ 4, the total digits of the dividend (344) are 3; and the total digit of the devisor (4) is 1.

The place to put the dividend will be 3 – 1 (division symbol) – 1 = 1.

Corresponding what we have learned earlier, 1 means to start putting the dividend from the ones place. The following picture showed that the dividend (344) was placed from the ones place on the abacus.

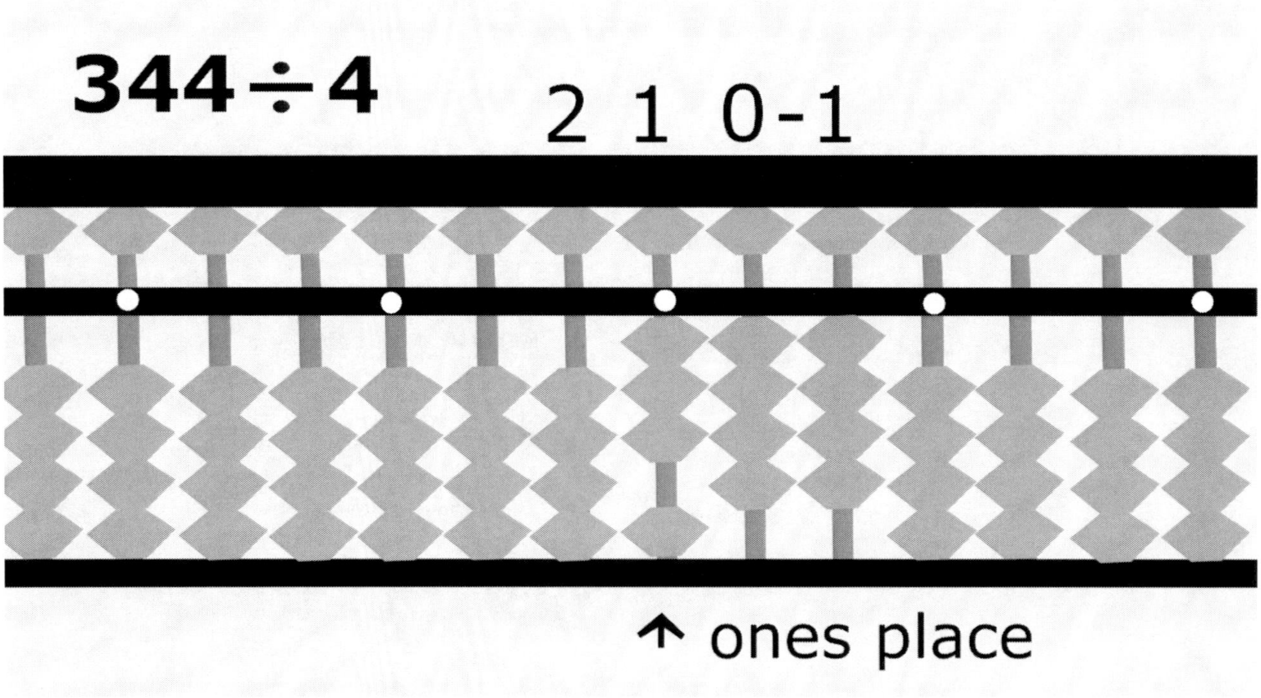

For the division question: 5687 ÷ 121, the total digits of the dividend (5687) are 4; and the total digit of the devisor (121) is 3.

The place to put the dividend will be 4 – 1 (division symbol) – 3 = 0. Corresponding what we have learned earlier, 0 means to start putting the dividend from tenth place. The following picture showed that the dividend (5687) was placed from the tenth place on the abacus.

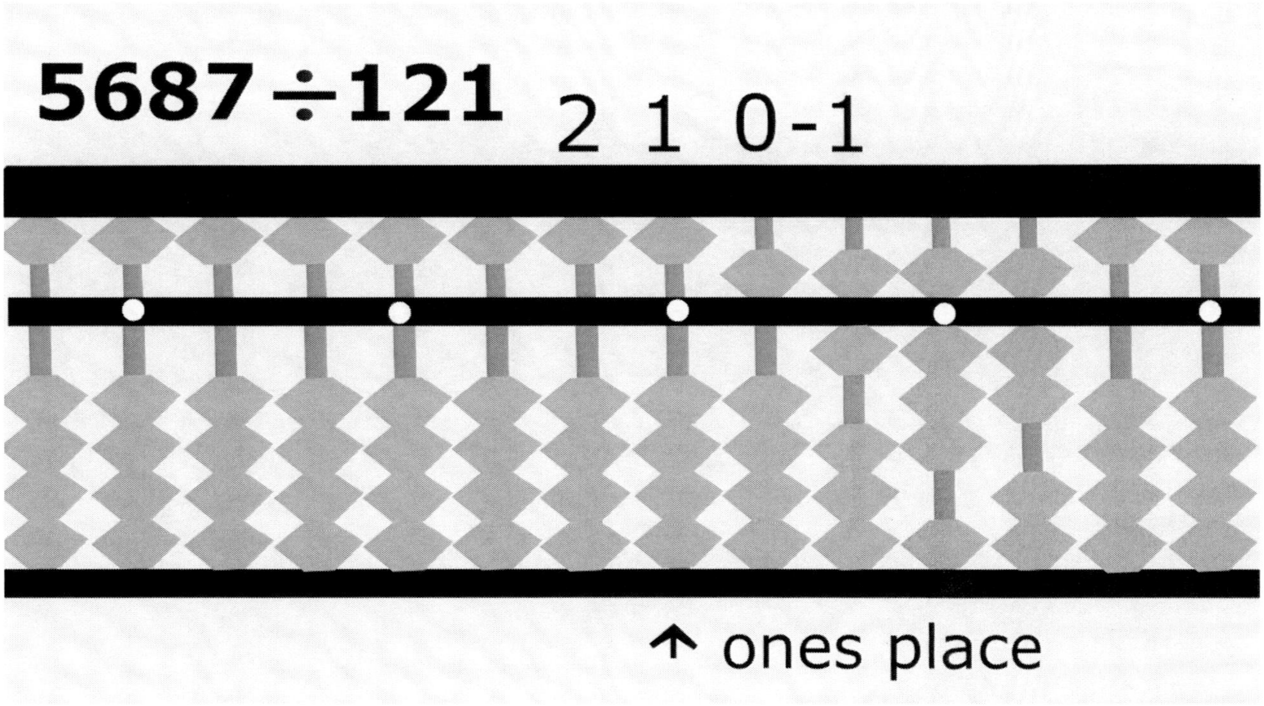

For the division question: 8358 ÷ 2, the total digits of the dividend (8358) are 4; and the total digit of the devisor (2) is 1.

The place to put the dividend will be 4 – 1 (division symbol) – 1 = 2.

Corresponding what we have learned earlier, 2 means to start putting the dividend from tens place. The following picture showed that the dividend (8358) was placed from the tens place on the abacus.

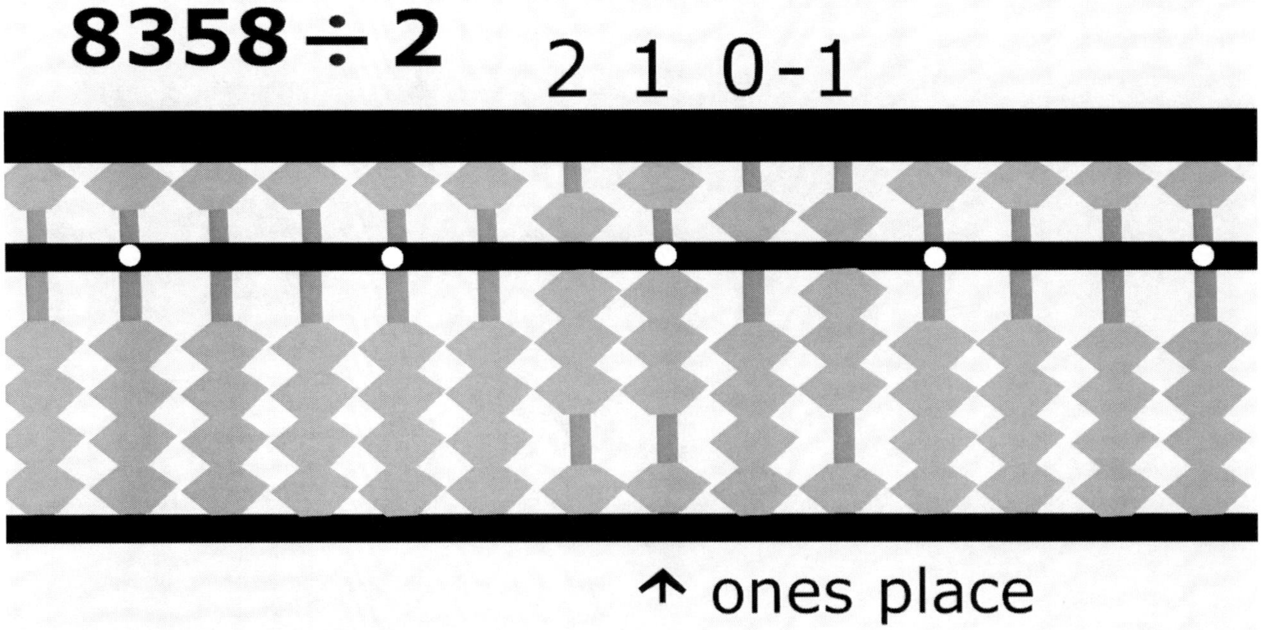

For the division question: 96 ÷ 32, the total digits of the dividend (96) are 2; and the total digit of the devisor (32) is 2.

The place to put the dividend will be 2 – 2 – 1 (÷) = -1. Corresponding what we have learned earlier, -1 means to start putting the dividend from hundredth place. The picture below showed that the dividend (96) was placed from the hundredth place on the abacus.

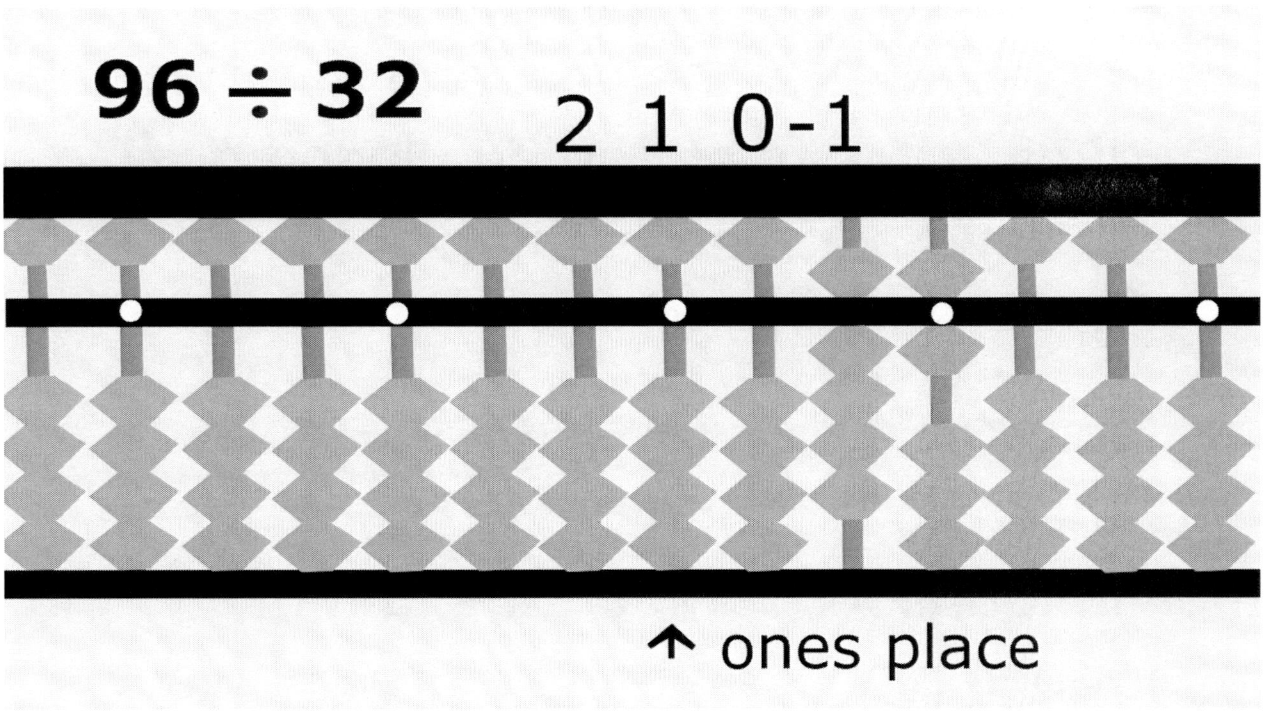

5. Quotient on the abacus

Once the position of the dividend has been settled, we can start the division operation, which is very similar to multiplication. However, there's another rule to keep in mind.

The differences between the first quotient number and the first dividend number determine the position of the division result on the abacus.

If the first quotient number is **large** than the first number of the dividend number, the first quotient number can be placed next to the dividend number as the picture on the next page showed.

However, If the first quotient number is **less** than the first number of the dividend, the first quotient number has to move to the left for one place, as the picture showed.

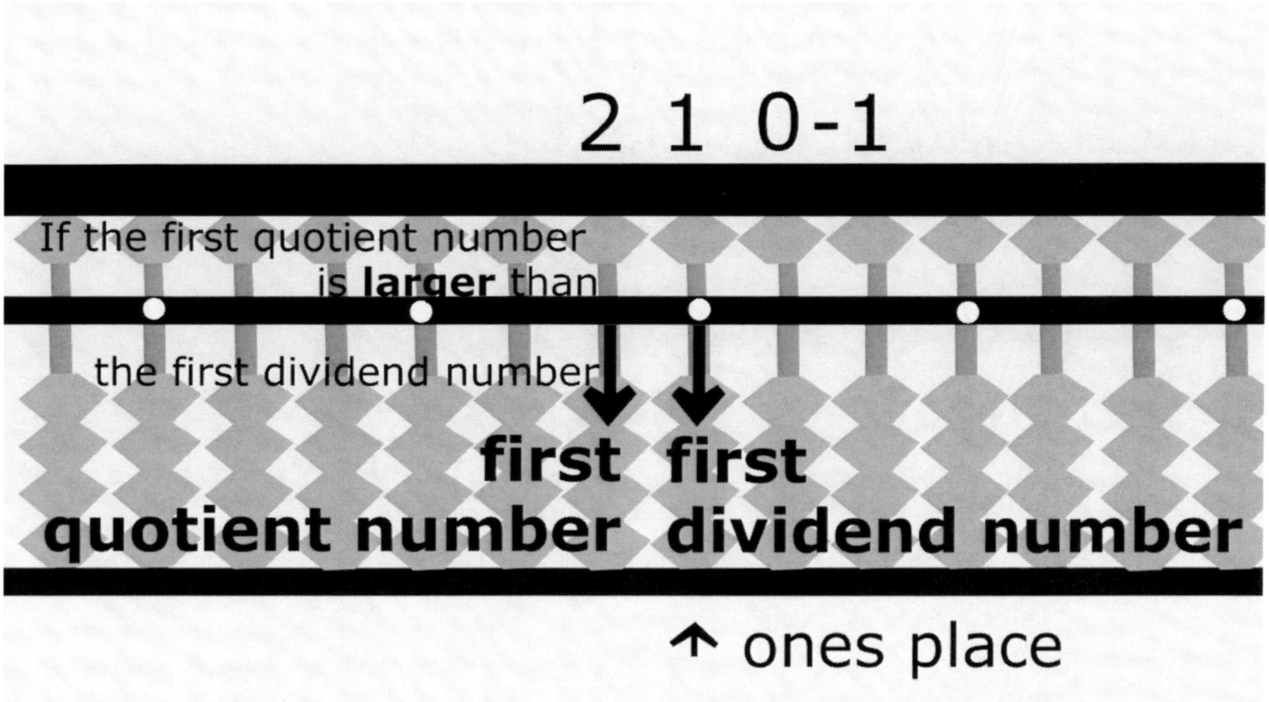
2 1 0-1
If the first quotient number
is **larger** than
the first dividend number
first quotient number
first dividend number
ones place

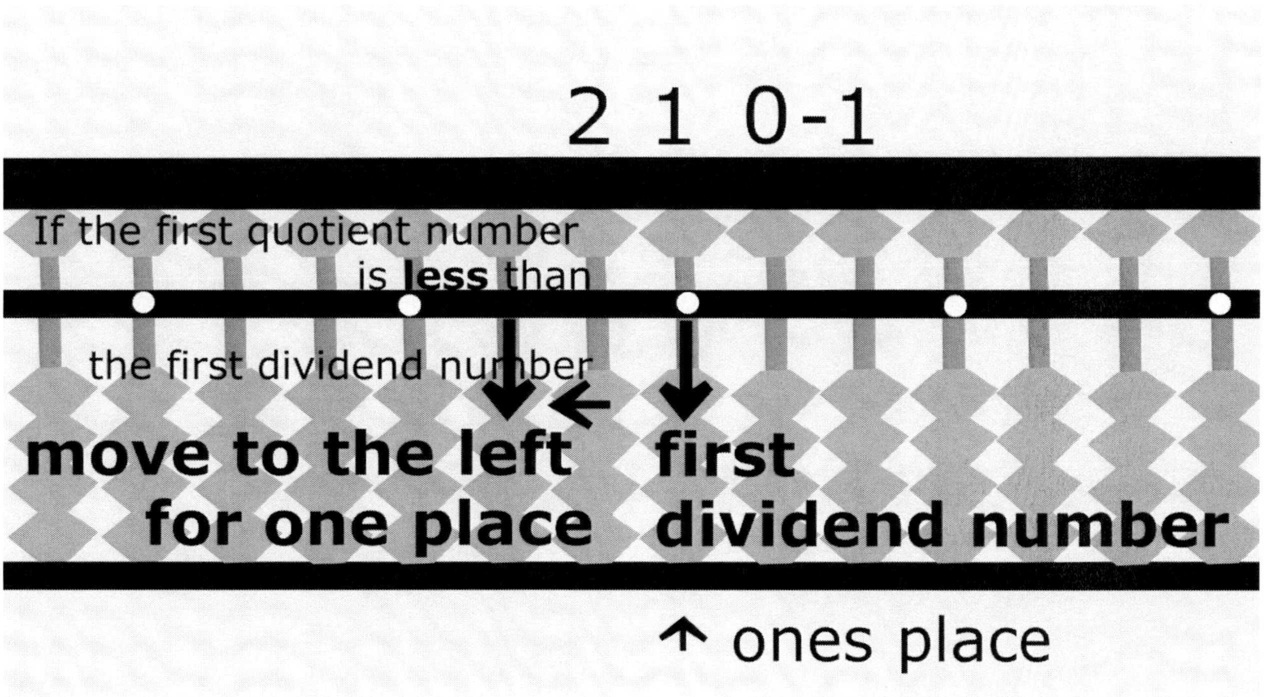
2 1 0-1
If the first quotient number
is **less** than
the first dividend number
move to the left for one place
first dividend number
ones place

For the sample question of 344 ÷ 4, the first quotient number is 8, which is the first closest quotient number for 34 ÷ 4.

8, the first quotient number, is larger than the first dividend number of 3, so the first quotient number of 8 is placed next to the first dividend number of 3, which is the tens place.

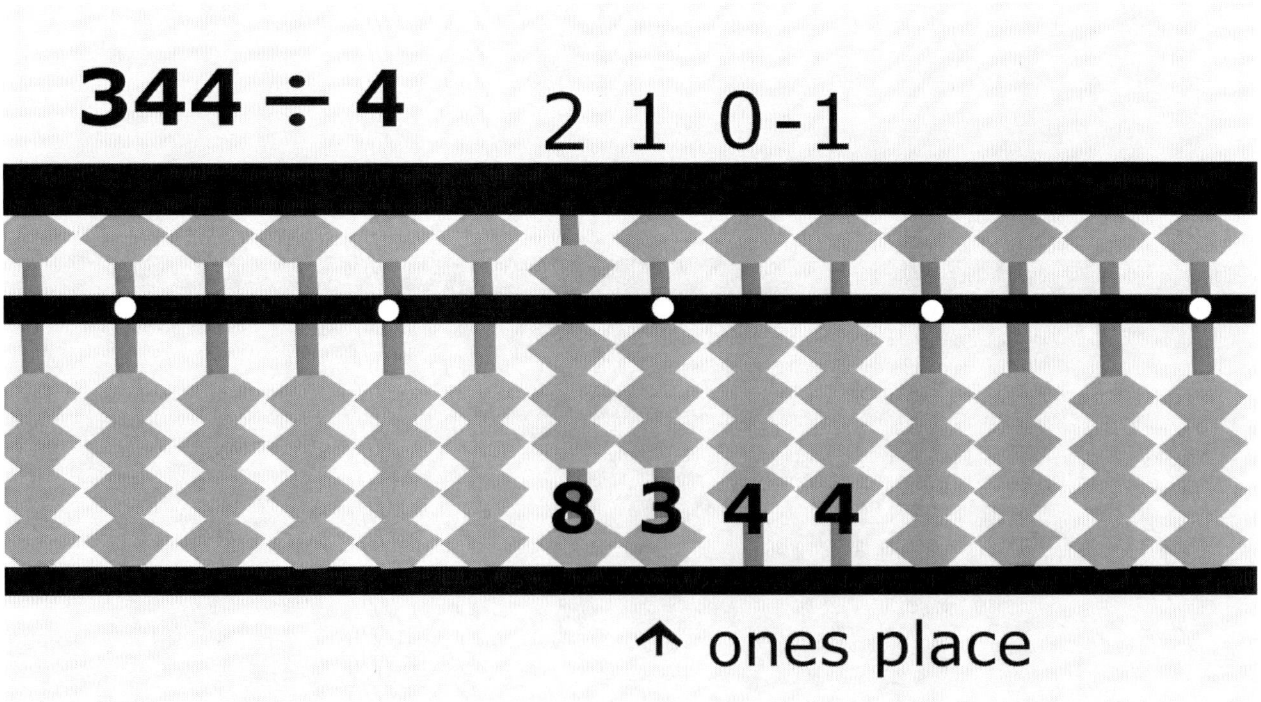

For the example question of 560 ÷ 4, the first quotient number is 1, which is the first closest quotient number for 5 ÷ 4.

1, the first quotient number, is less than the first dividend number of 5, so the number of 1 needs to move to the left for one place. The picture below showed that the first quotient number of 1 is put at the hundreds place, one place apart from the first dividend number of 5.

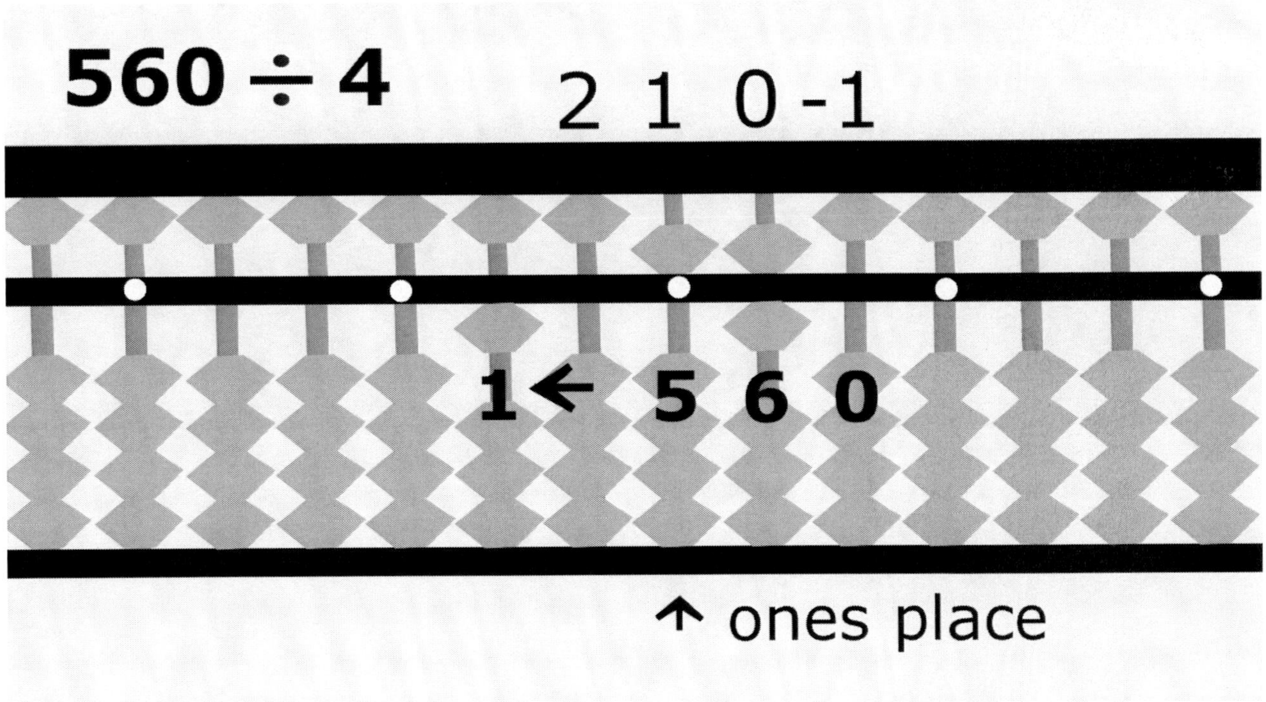

For the sample question of 408 ÷ 8, the first quotient number is 5, which is the first closest quotient number for 40 ÷ 8.

5, the first quotient number, is larger than the first dividend number of 4, so the first quotient number of 5 is placed next to the first dividend number of 4, which is the tens place.

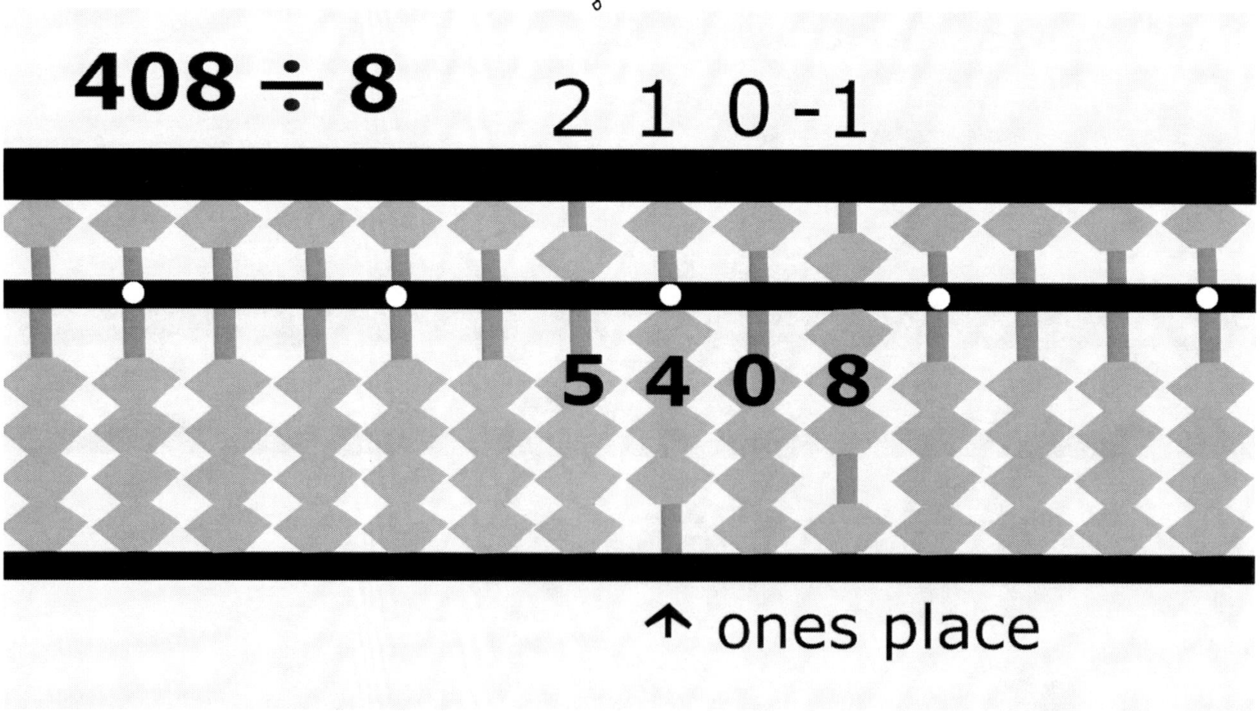

For the example question of 5687 ÷ 121, the first quotient number is 4, which is the first closest quotient number for 568 ÷ 121.

4, the first quotient number, is less than the first dividend number of 5, so the number of 4 needs to move to the left for one place. The picture below showed that the first quotient number of 4 is put at the tens place, one place apart from the first dividend number of 5.

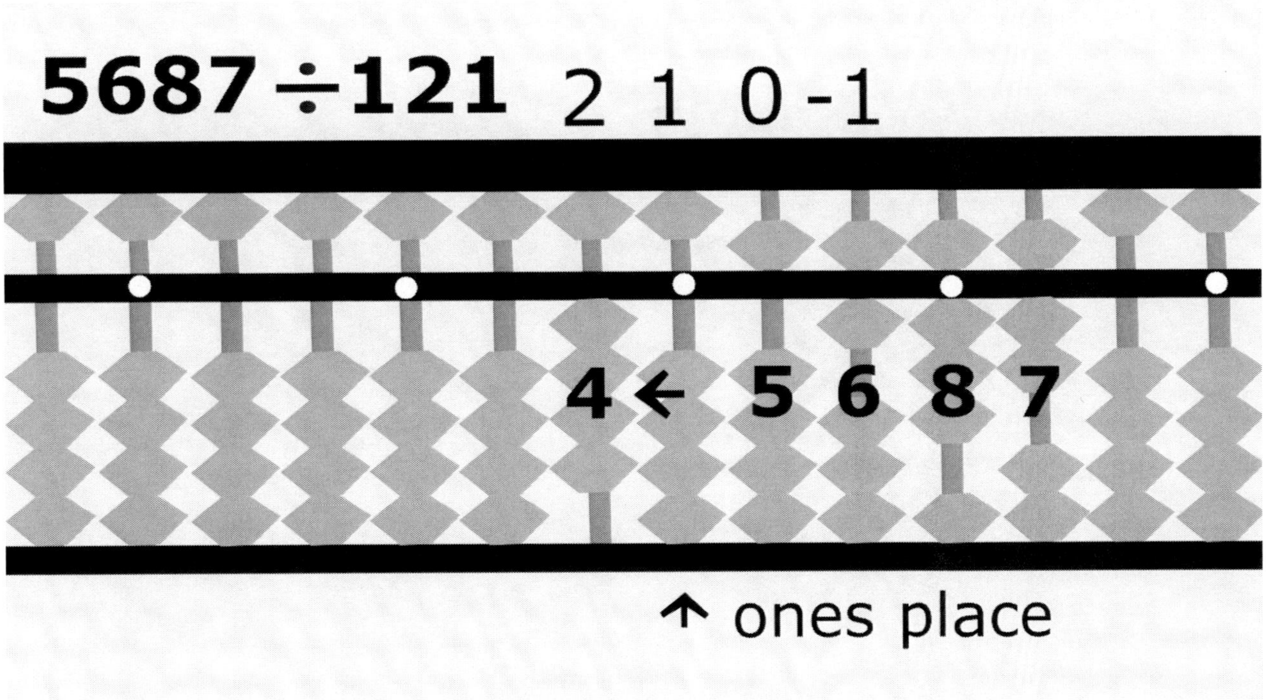

6. Division Sequence

When solving multiplication problems, we multiply numbers one by one as shown in the pictures below.

Multiplication

Multiplication

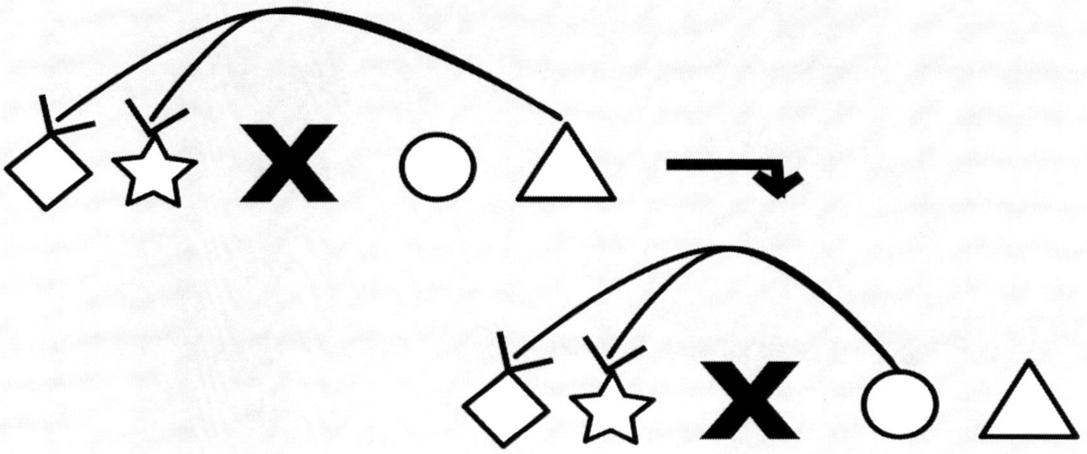

For the division operation, it's important to treat the devisor, one digit, two digits or three digits as one subject while solving division questions. You can have the dividend be divided by each factor of the devisor but people normally don't do that.

The following pictures showed the Division Sequence while solving division questions. At each step, we need to take the remainders into calculation if applicable.

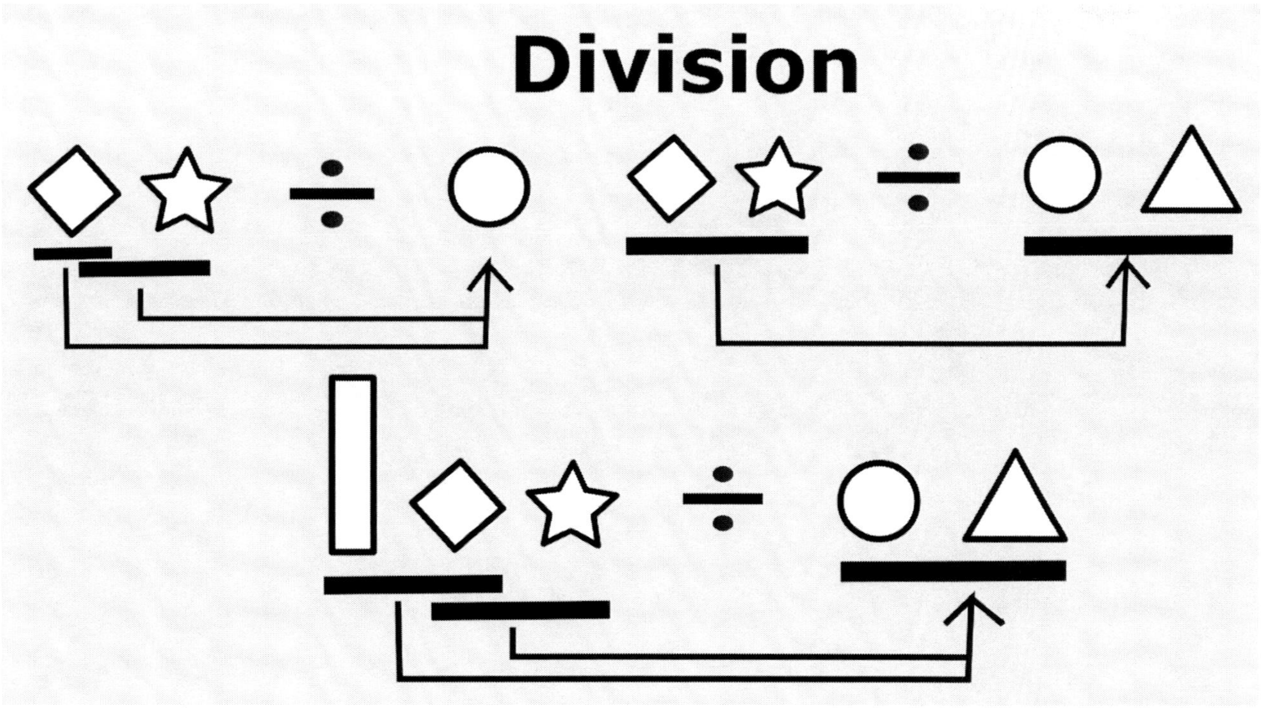

7. two digits divided by one digit

Dividend on the abacus: 2 digits of dividend – 1 division symbol – 1 digit of divisor = dividend position on the abacus. The abacus position for 48 is 0, on the tenth place.

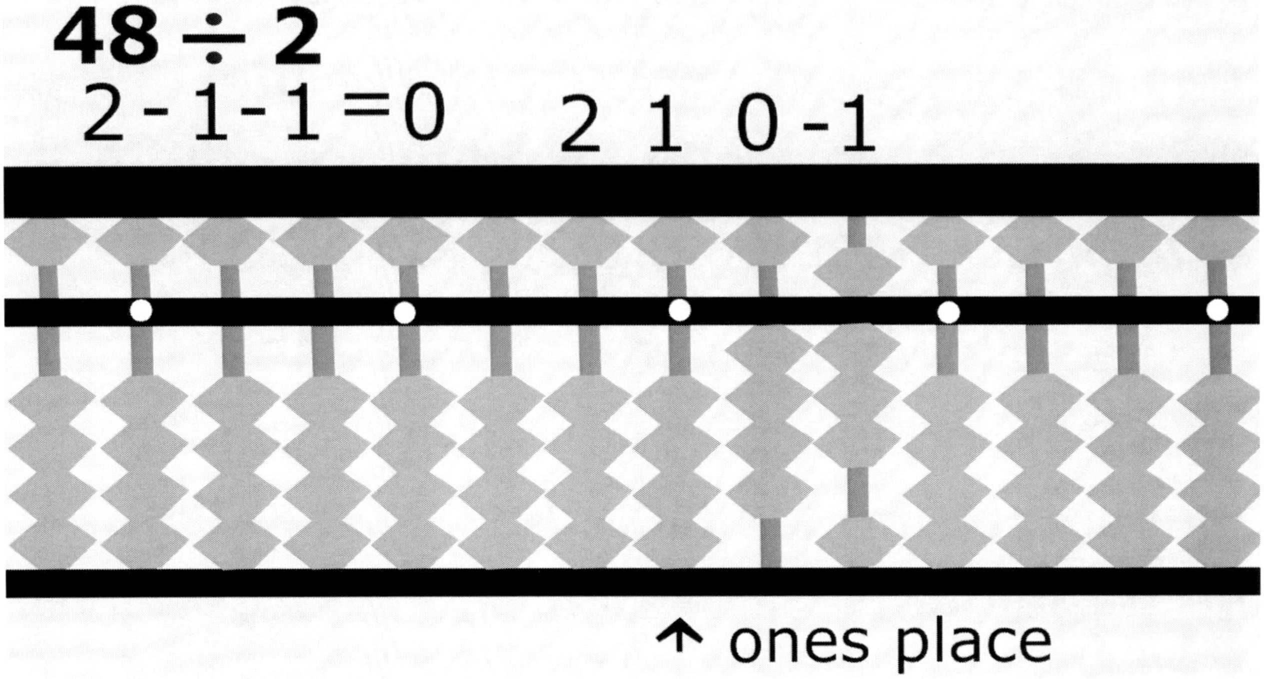

Quotient on the abacus: 4 ÷ 2 = 2, 2 < 4, so the quotient, 2, moves to the left for one place.

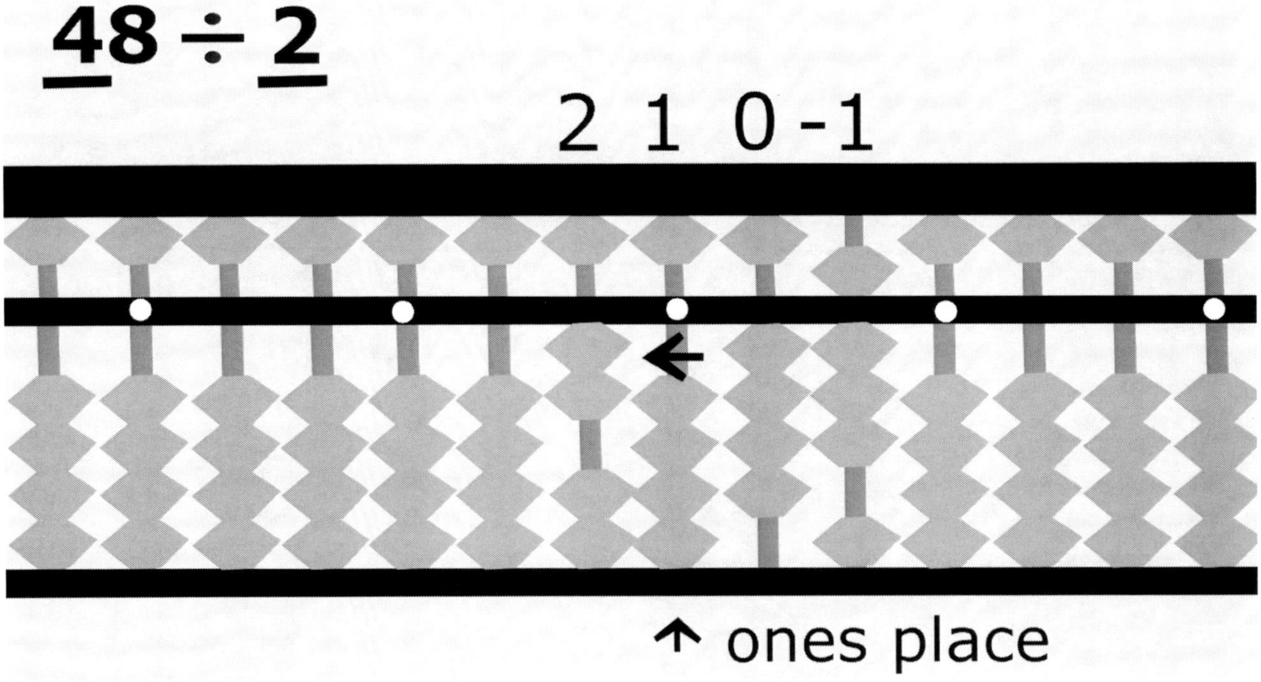

Quotient on the abacus: 4 ÷ 2 remainder is 0, the next step would be 08 ÷ 2.

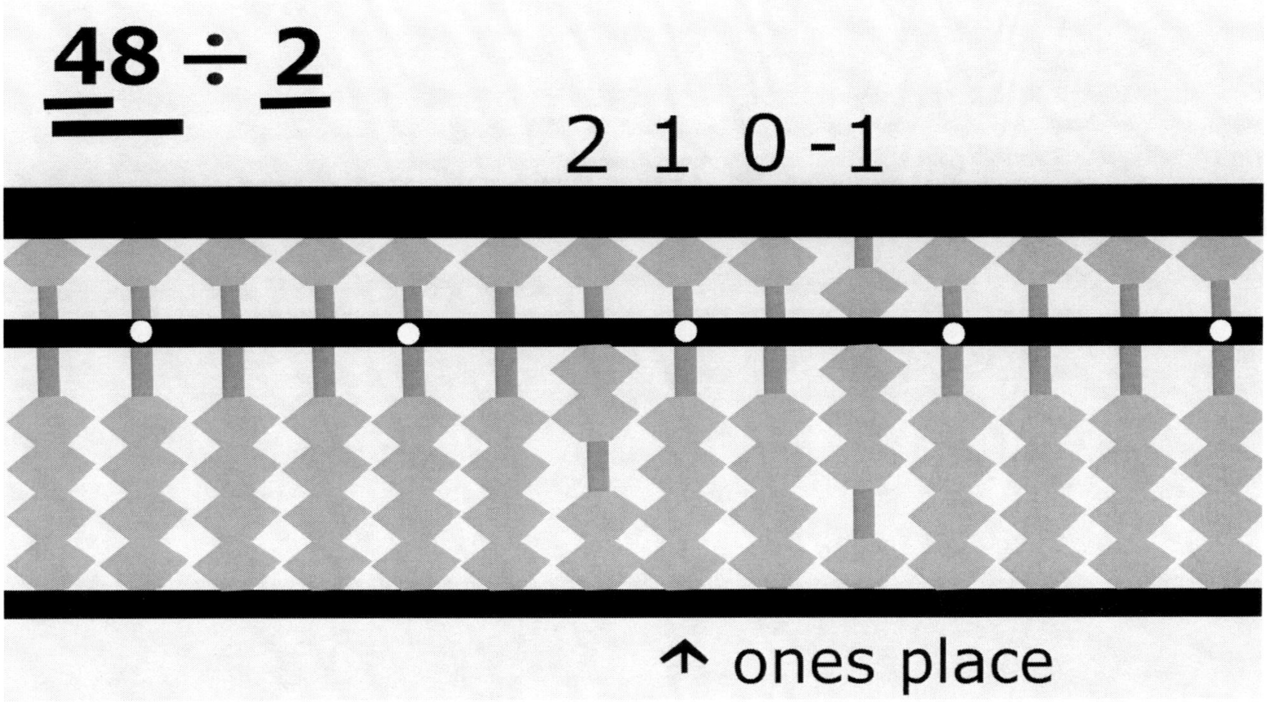

Quotient on the abacus: 48 ÷ 2 = 24

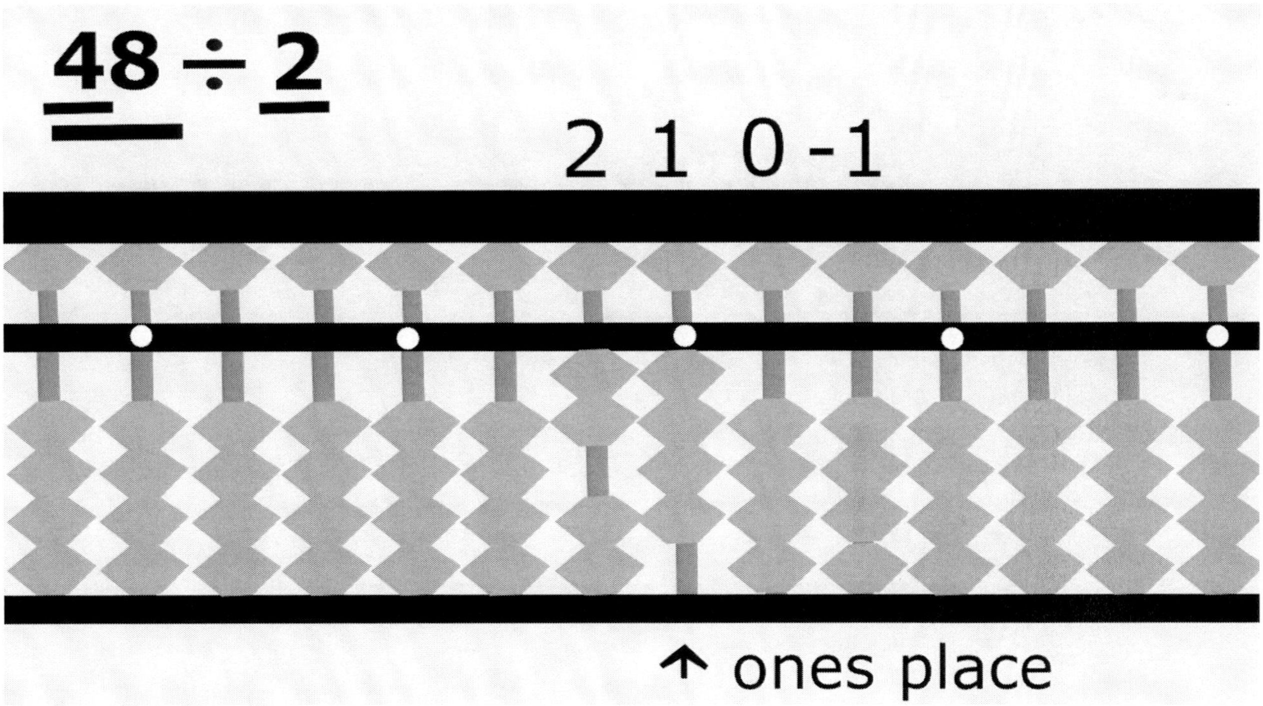

Dividend on the abacus: 2 digits of dividend – 1 division symbol – 1 digit of divisor = dividend position on the abacus. The abacus position for 95 is 0, on the tenth place.

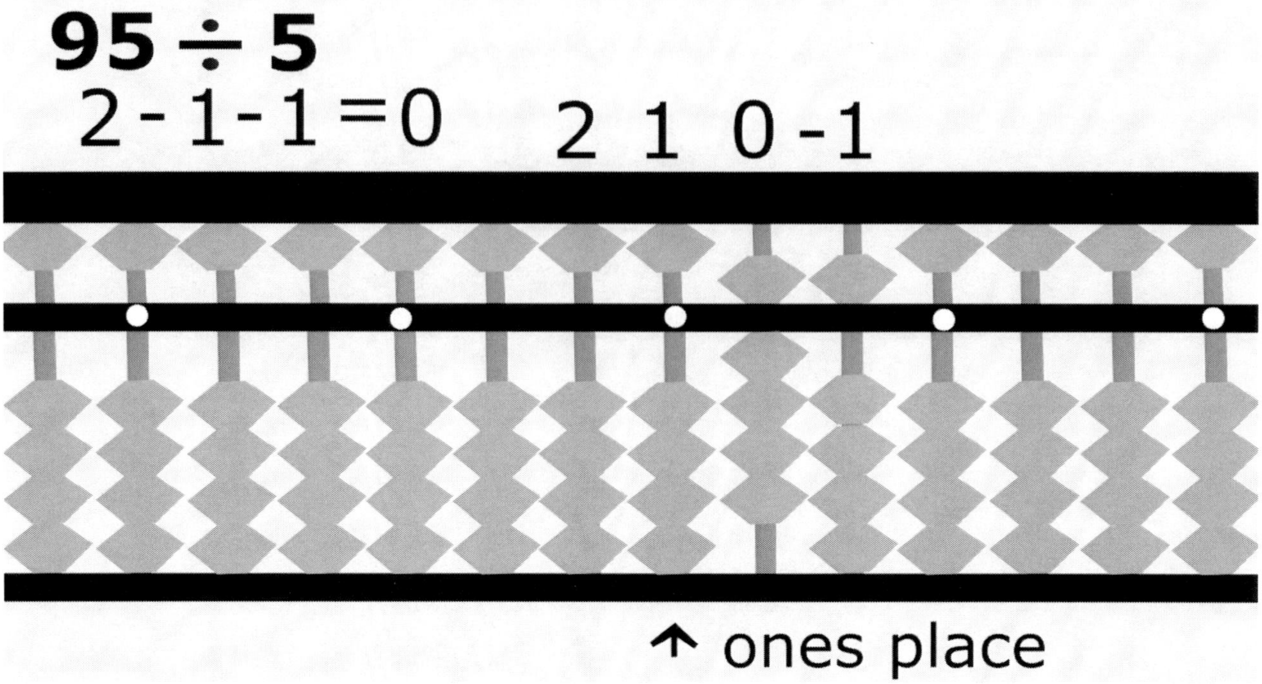

Quotient on the abacus: 9 ÷ 5 = 1 and remainder, 1 < 9, so the quotient,1, moves to the left for one place.

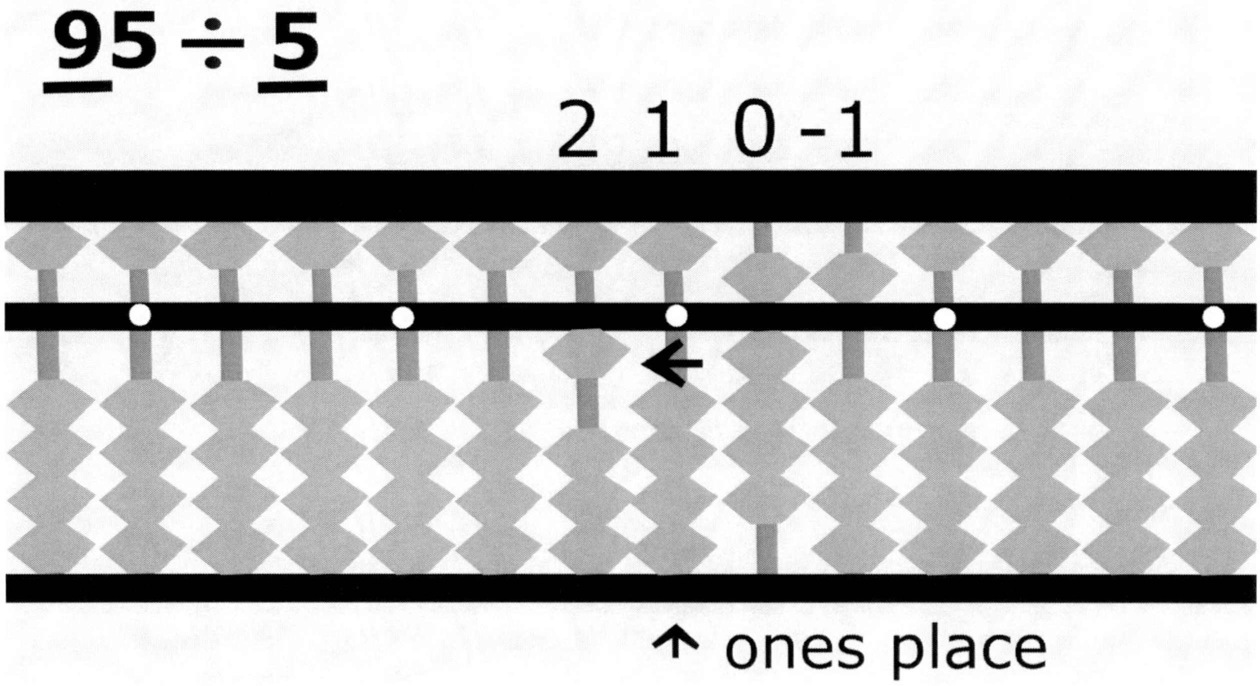

Quotient on the abacus: 9 ÷ 5 remainder is 4, the next step would be 45 ÷ 5.

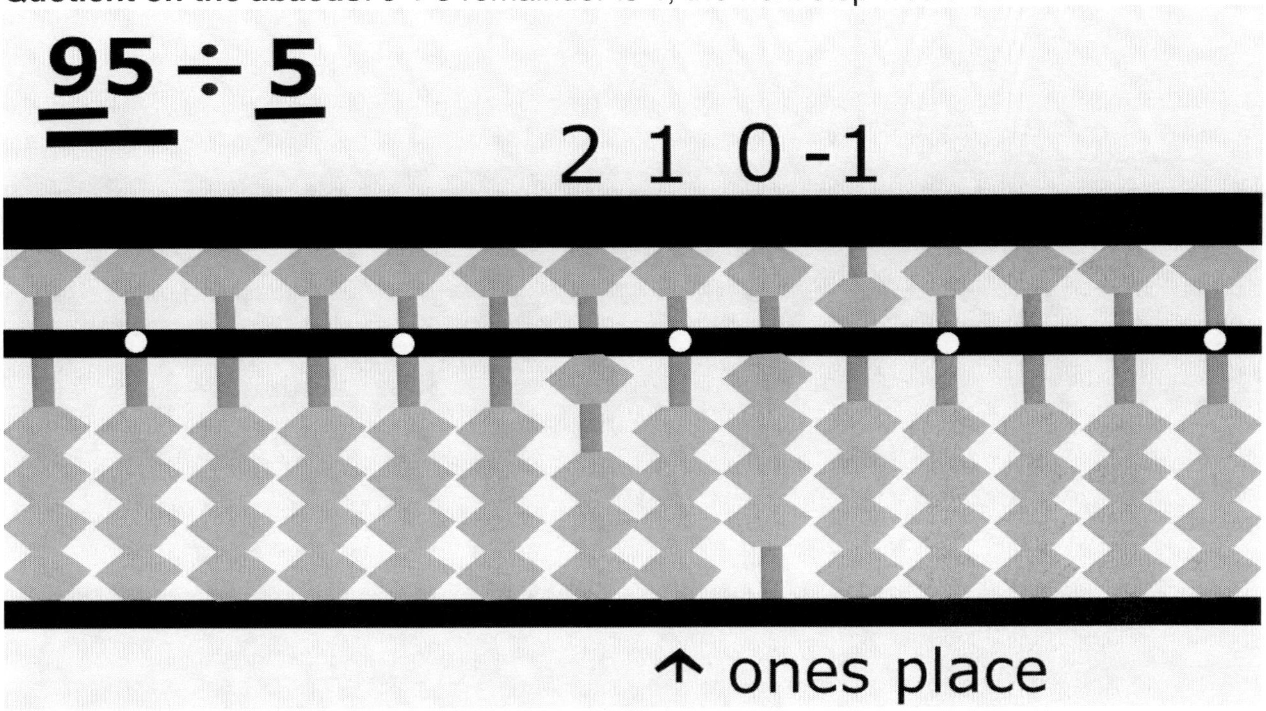

Quotient on the abacus: 95 ÷ 5 = 19

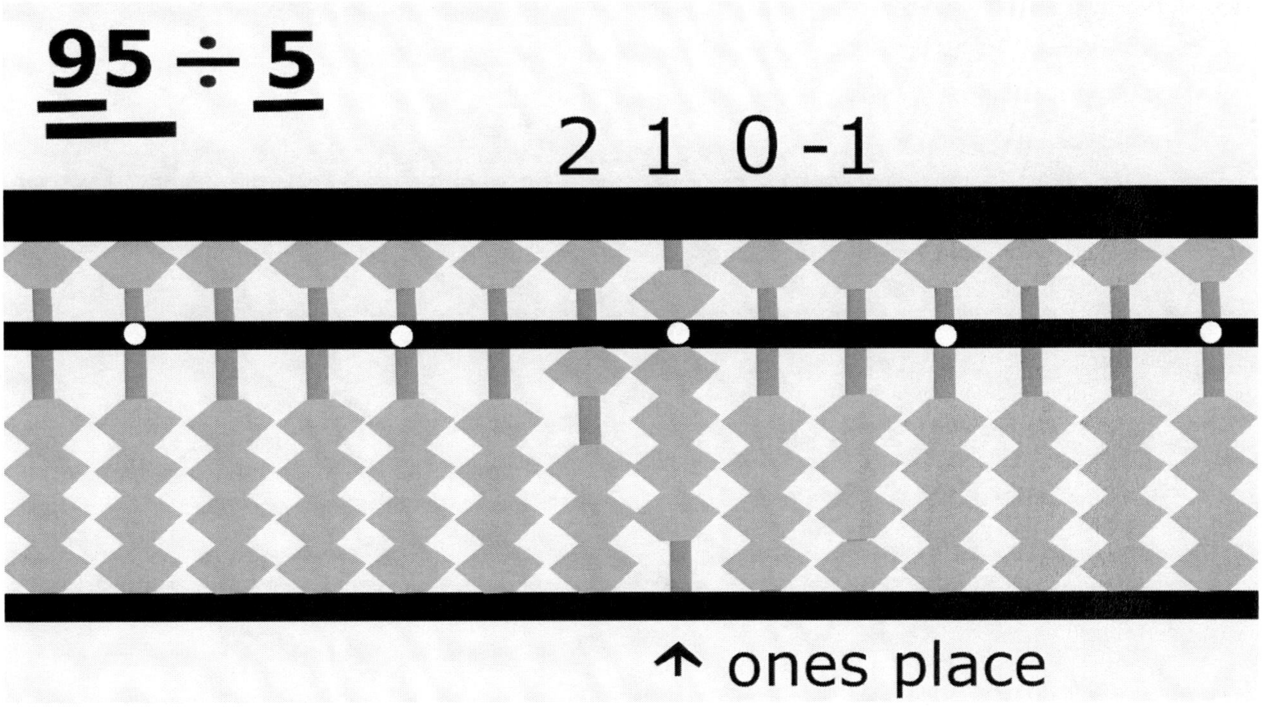

Dividend on the abacus: 2 digits of dividend - 1 division symbol – 1 digit of divisor = dividend position on the abacus. The abacus position for 56 is 0, on the tenth place.

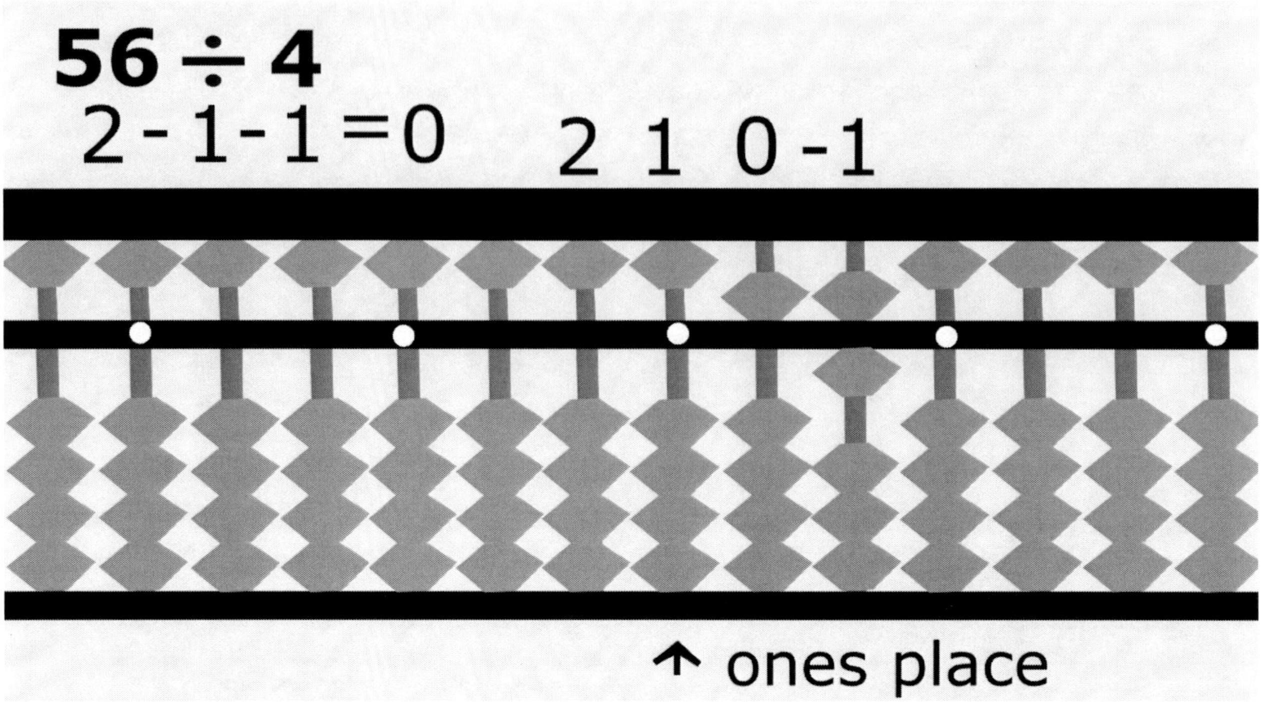

Quotient on the abacus: 5 ÷ 4 = 1 and remainder, 1 < 5, so the quotient, 1, moves to the left for one place.

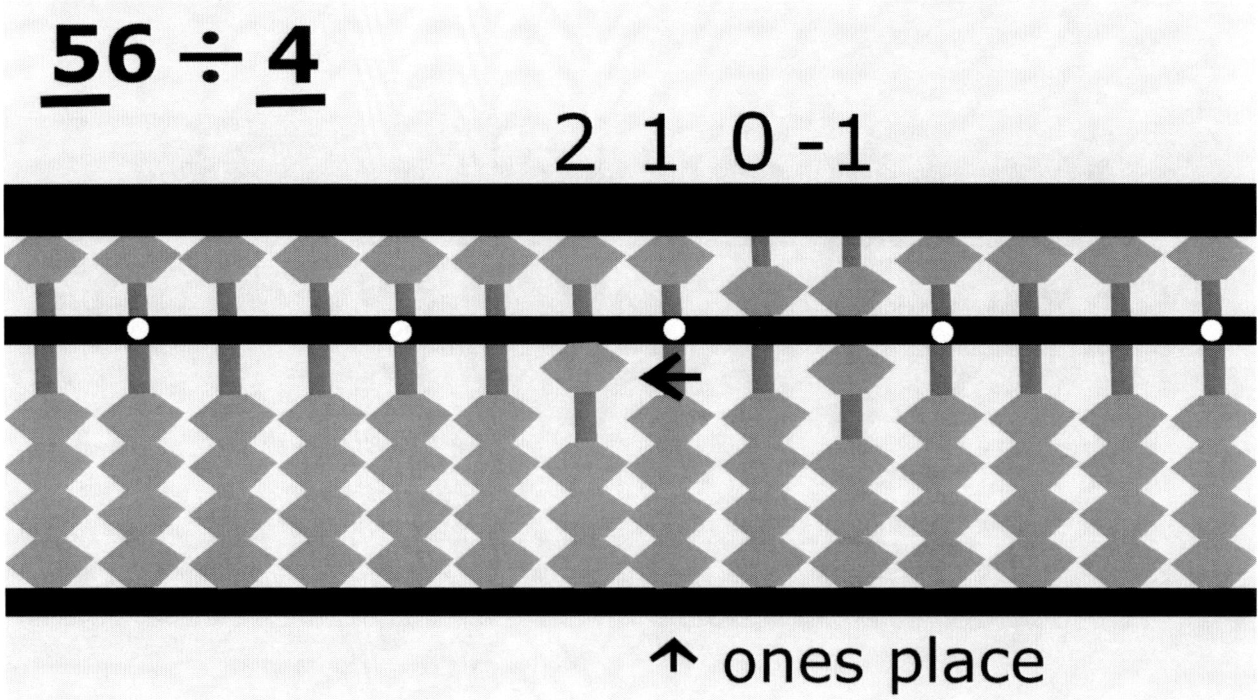

Quotient on the abacus: 5 ÷ 4 remainder is 1, the next step would be 16 ÷ 4.

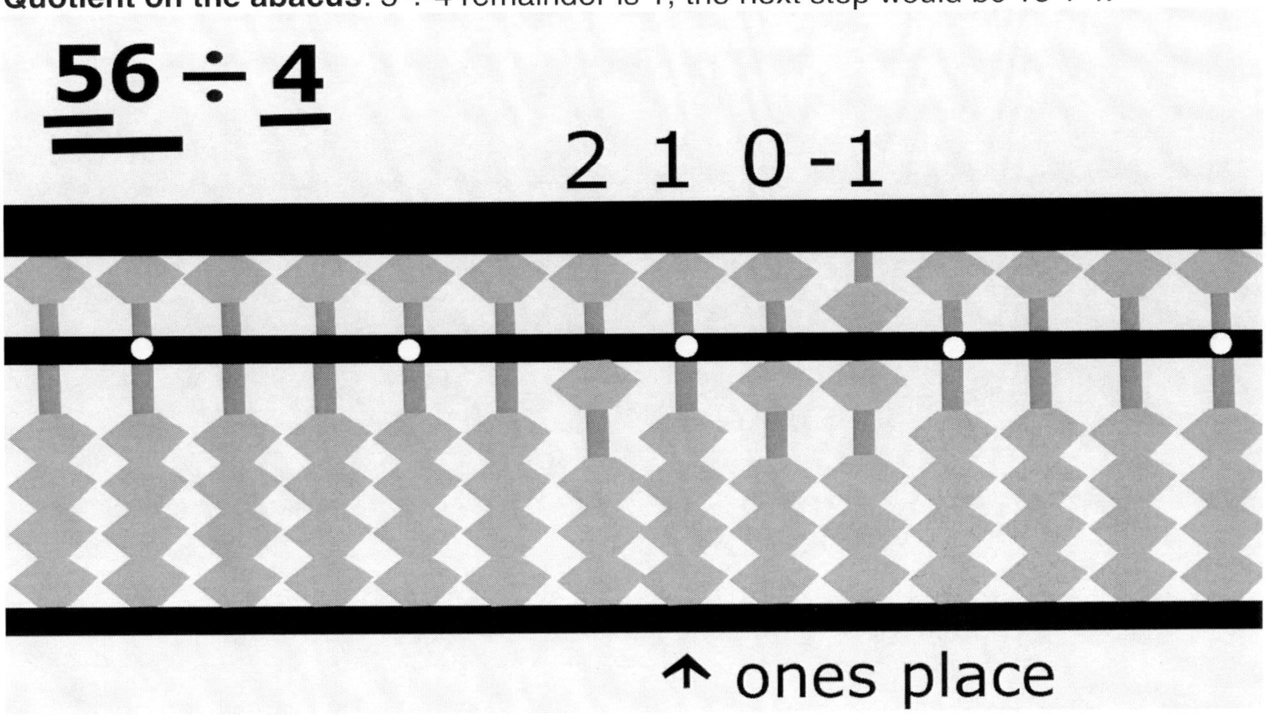

Quotient on the abacus: 56 ÷ 4 = 14

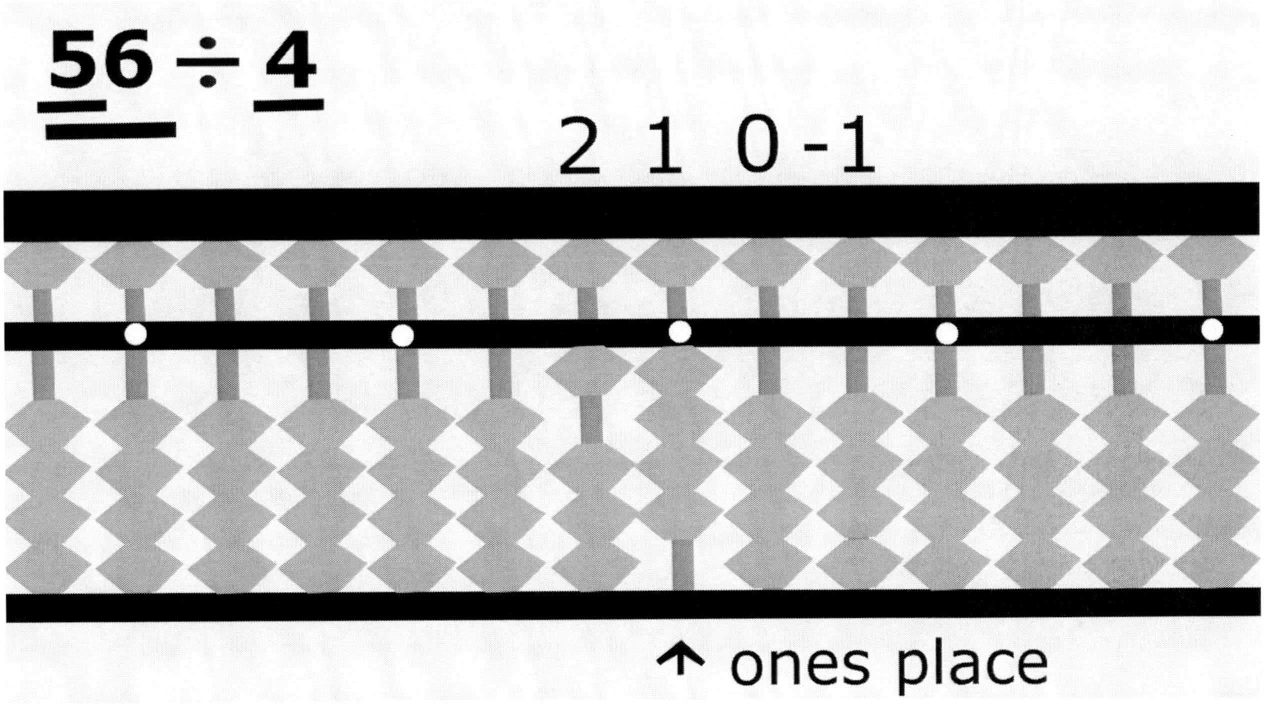

Practice for 2 digits divided by 1 digit

1) 66 ÷ 6 =	15) 52 ÷ 4 =
2) 96 ÷ 6 =	16) 87 ÷ 3 =
3) 65 ÷ 5 =	17) 85 ÷ 5 =
4) 81 ÷ 3 =	18) 99 ÷ 9 =
5) 92 ÷ 4 =	19) 69 ÷ 3 =
6) 72 ÷ 4 =	20) 94 ÷ 2 =
7) 65 ÷ 5 =	21) 84 ÷ 6 =
8) 75 ÷ 3 =	22) 78 ÷ 3 =
9) 68 ÷ 4 =	23) 80 ÷ 8 =
10) 76 ÷ 4 =	24) 77 ÷ 7 =
11) 88 ÷ 2 =	25) 30 ÷ 3 =
12) 78 ÷ 6 =	26) 96 ÷ 8 =
13) 56 ÷ 2 =	27) 96 ÷ 4 =
14) 45 ÷ 3 =	28) 66 ÷ 3 =

Albert Einstein:

Pure mathematics is, in its way, the poetry of logical ideas.

8. three digits divided by one digit

Dividend on the abacus: 3 digits of dividend – 1 division symbol – 1 digit of divisor = dividend position on the abacus. The abacus position for 123 is 1, on the ones place.

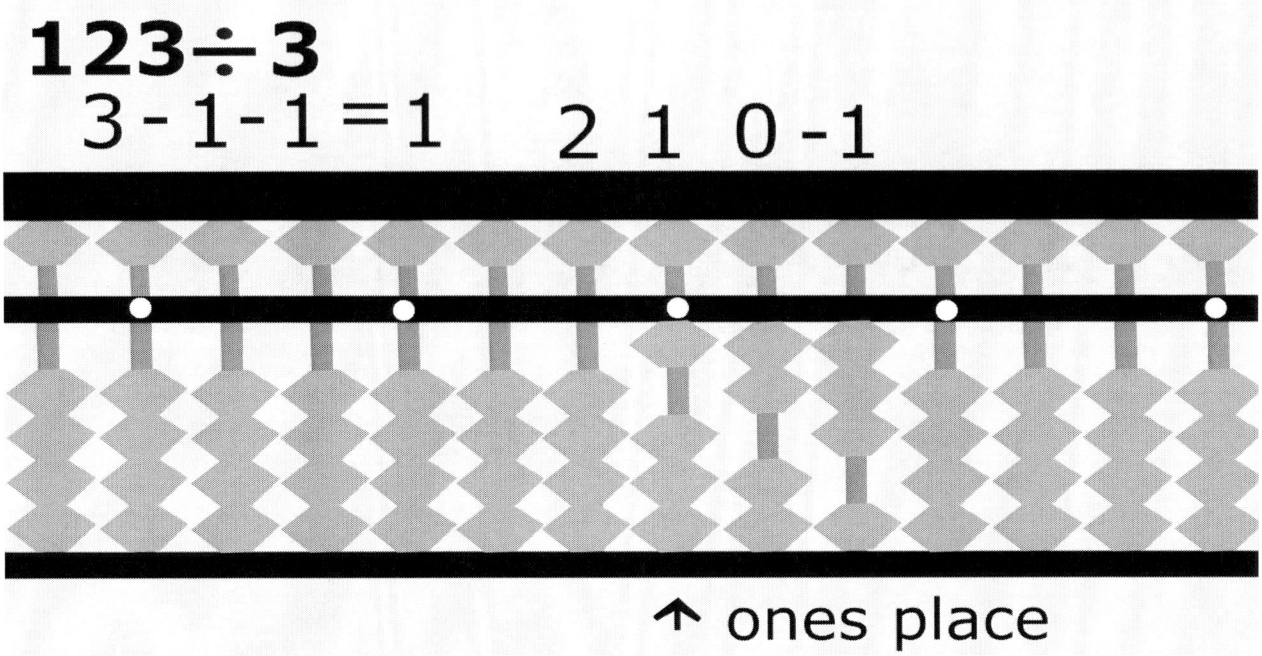

Quotient on the abacus: 1 cannot be divided by 3. 12 ÷ 3 = 4, 4 > 1, so the quotient, 4 does not need to move to the left for one place.

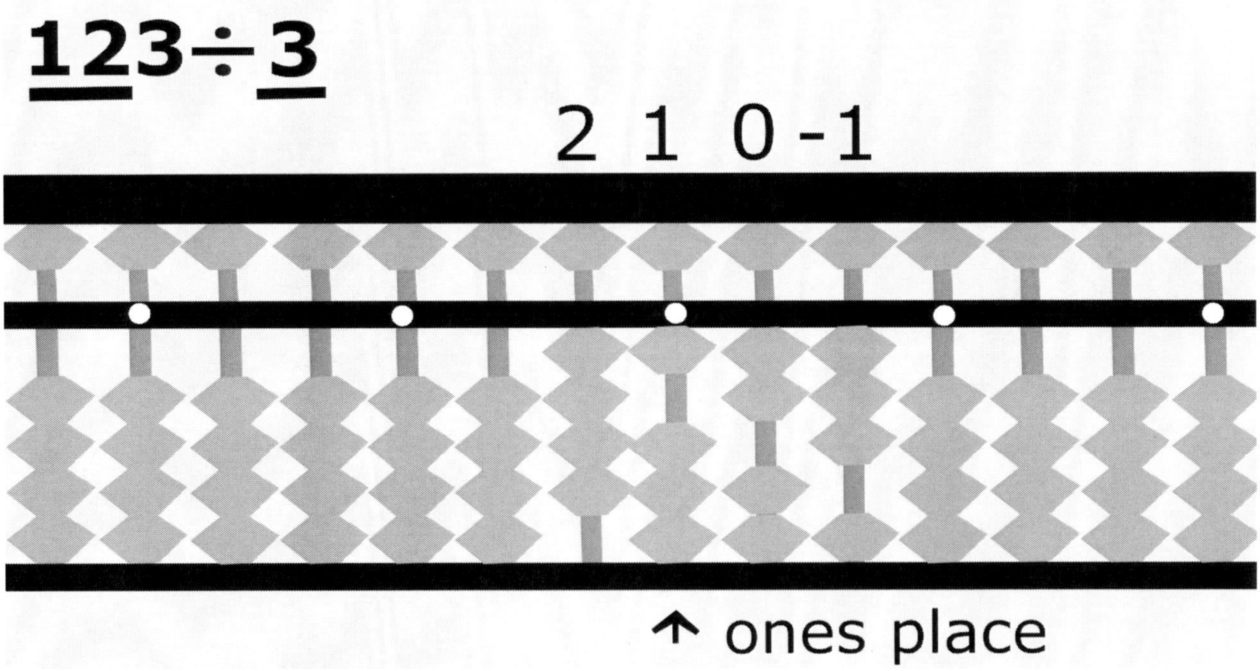

Quotient on the abacus: 12 ÷ 3 remainder is 0, the next step would be 03 ÷ 3.

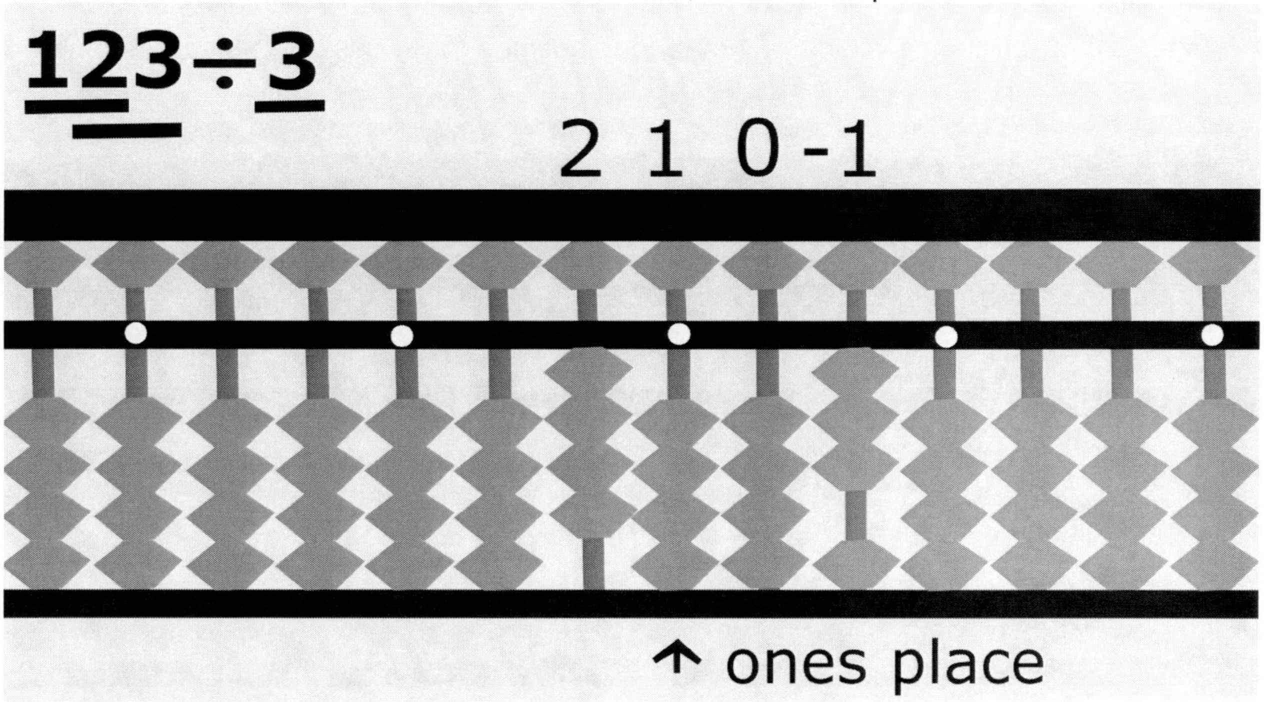

Quotient on the abacus: 123 ÷ 3 = 41

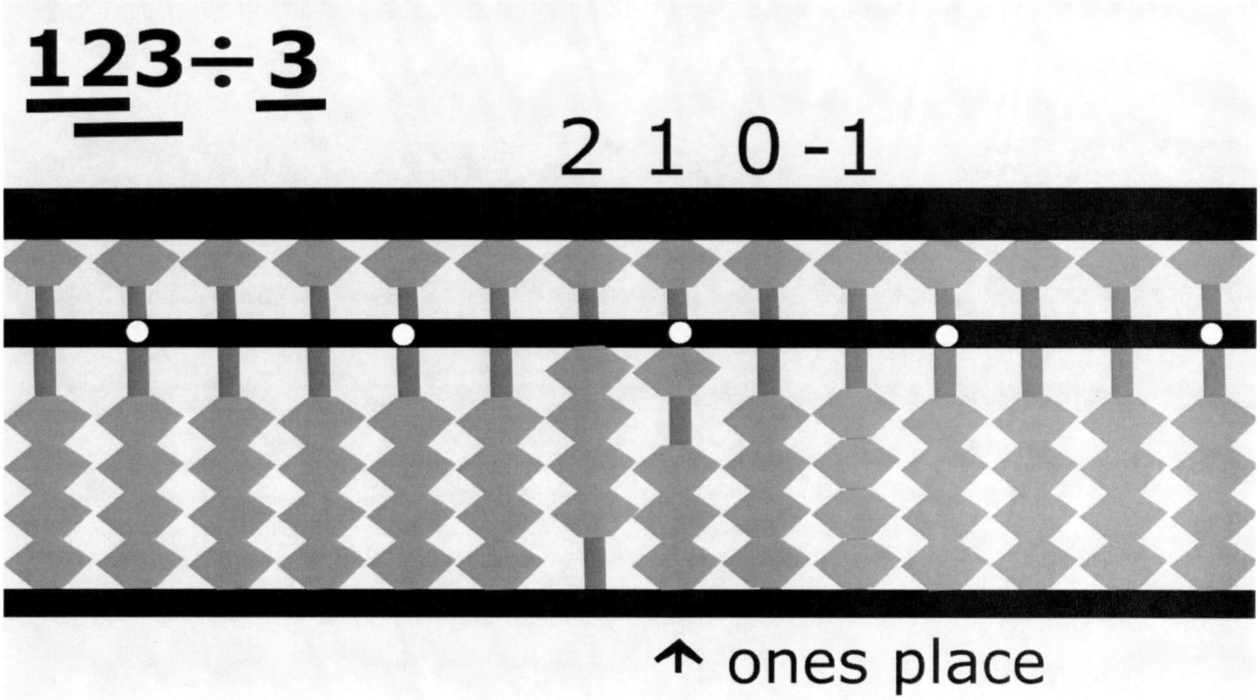

Dividend on the abacus: 3 digits of dividend – 1 division symbol – 1 digit of divisor = dividend position on the abacus. The abacus position for 258 is 1, on the ones place.

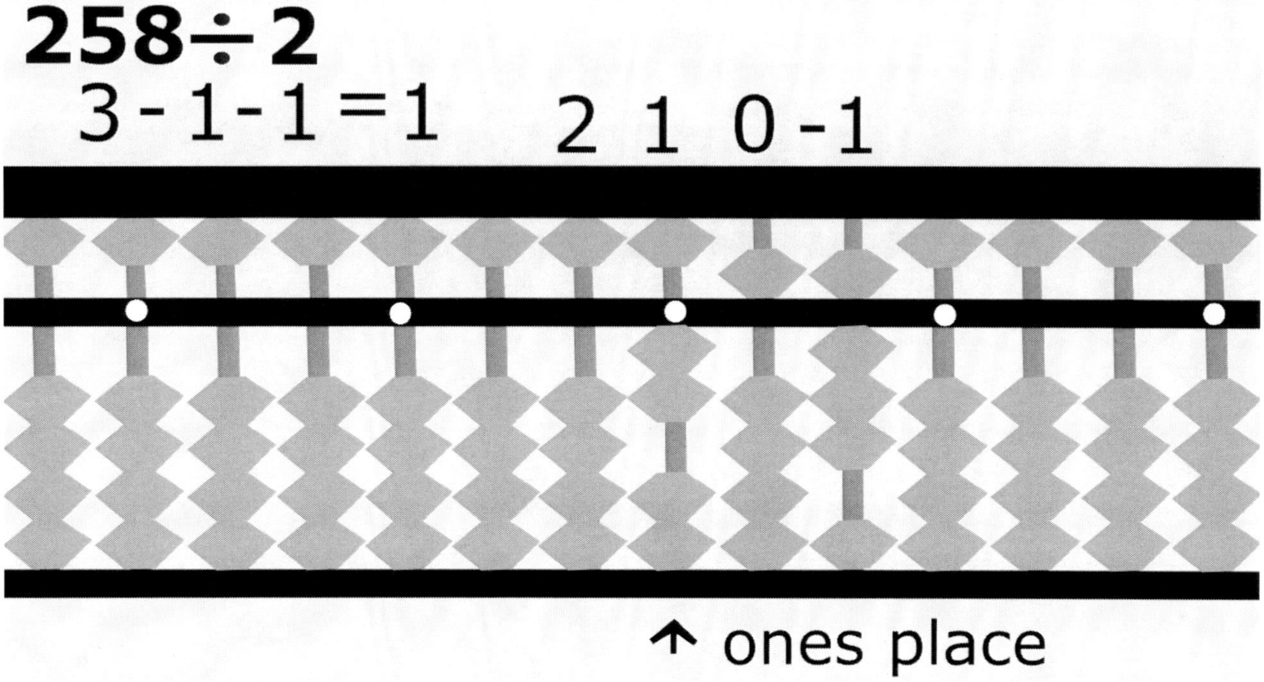

Quotient on the abacus: 2 ÷ 2 = 1, 1 < 2, so the quotient,1, moves to the left for one place.

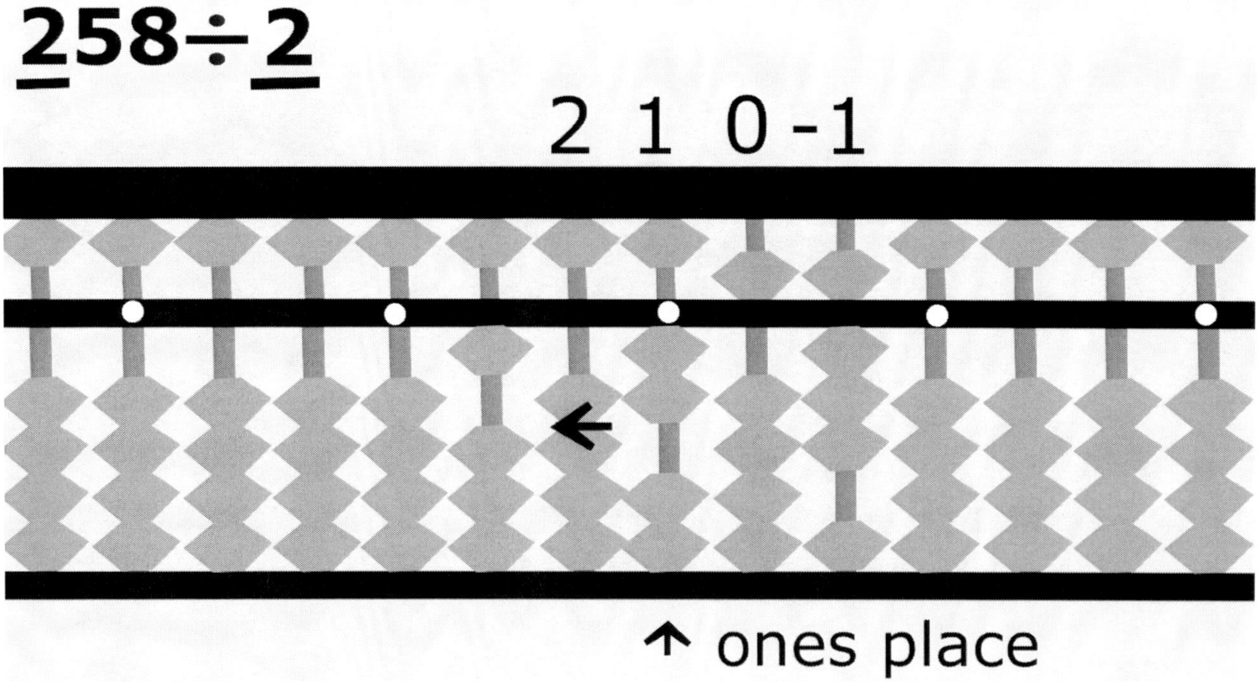

Quotient on the abacus: 2 ÷ 2 remainder is 0, the next step would be 05 ÷ 2

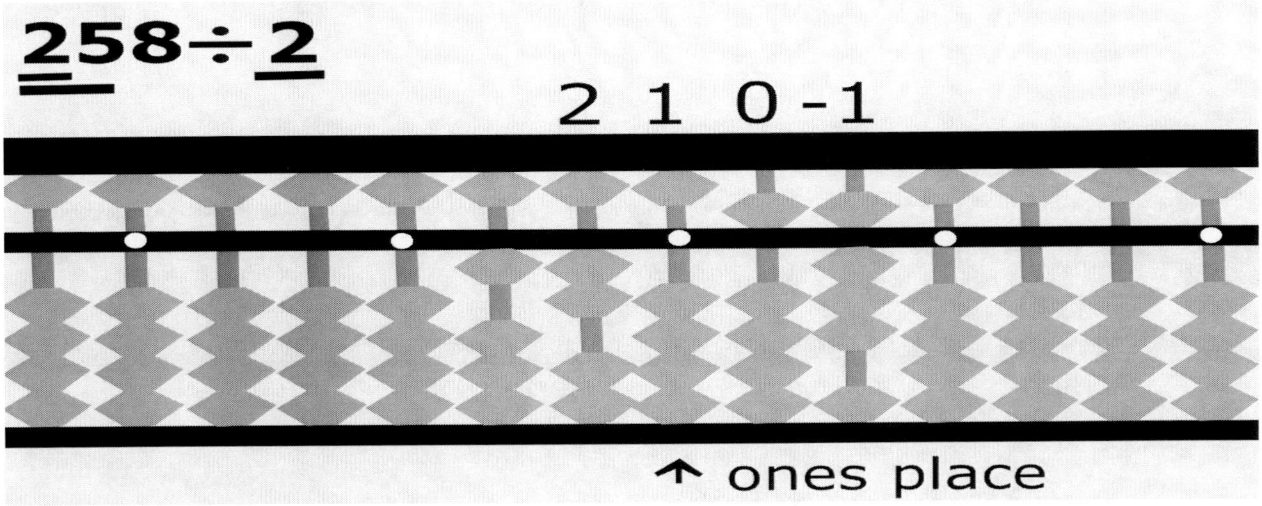

Quotient on the abacus: 05 ÷ 2 remainder is 1, the next step would be 18 ÷ 2

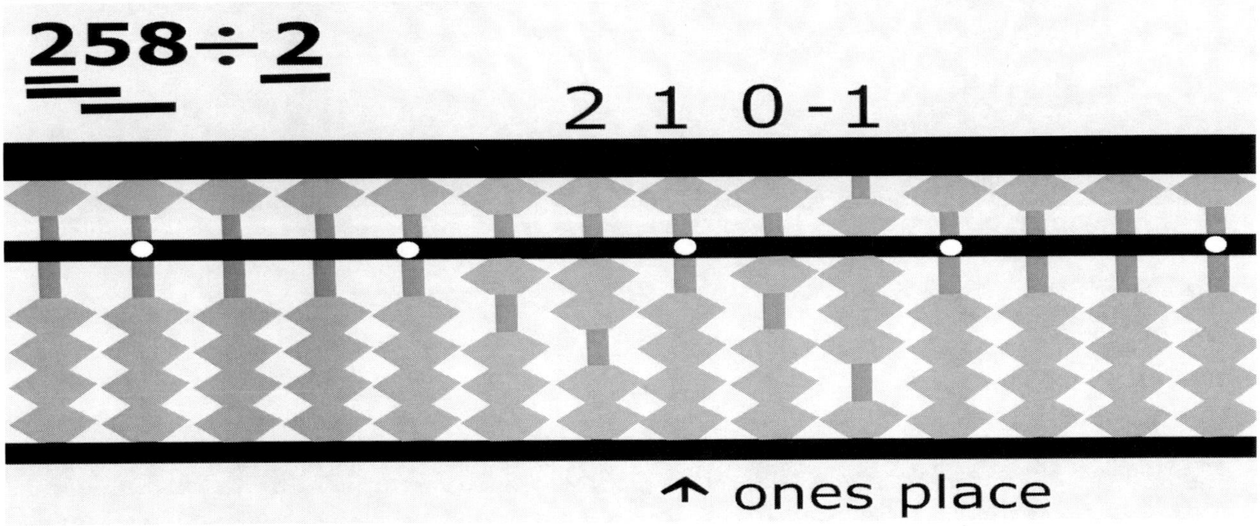

Quotient on the abacus: 258 ÷ 2 = 129

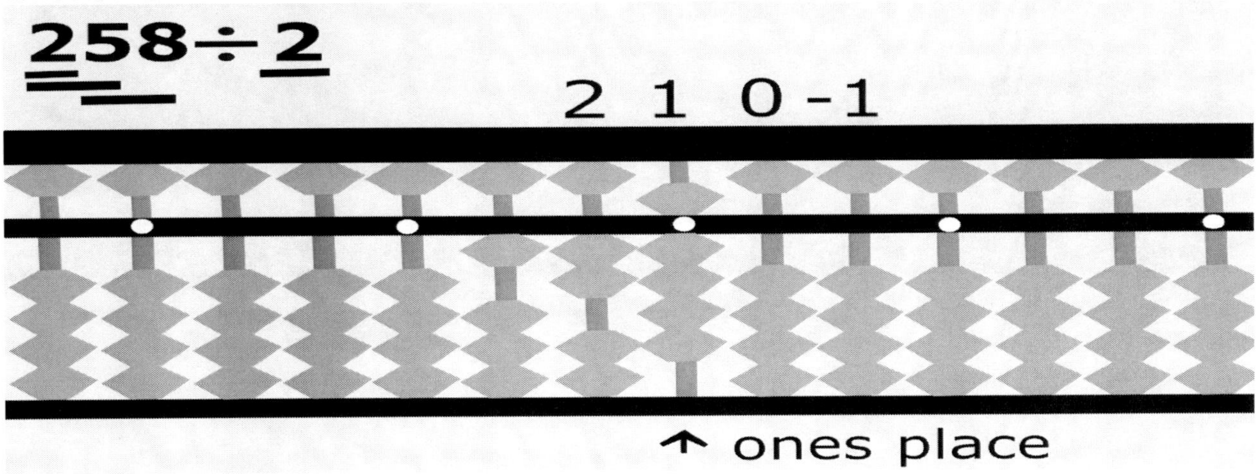

Dividend on the abacus: 3 digits of dividend – 1 division symbol – 1 digit of divisor = dividend position on the abacus. The abacus position for 861 is 1, on the ones place.

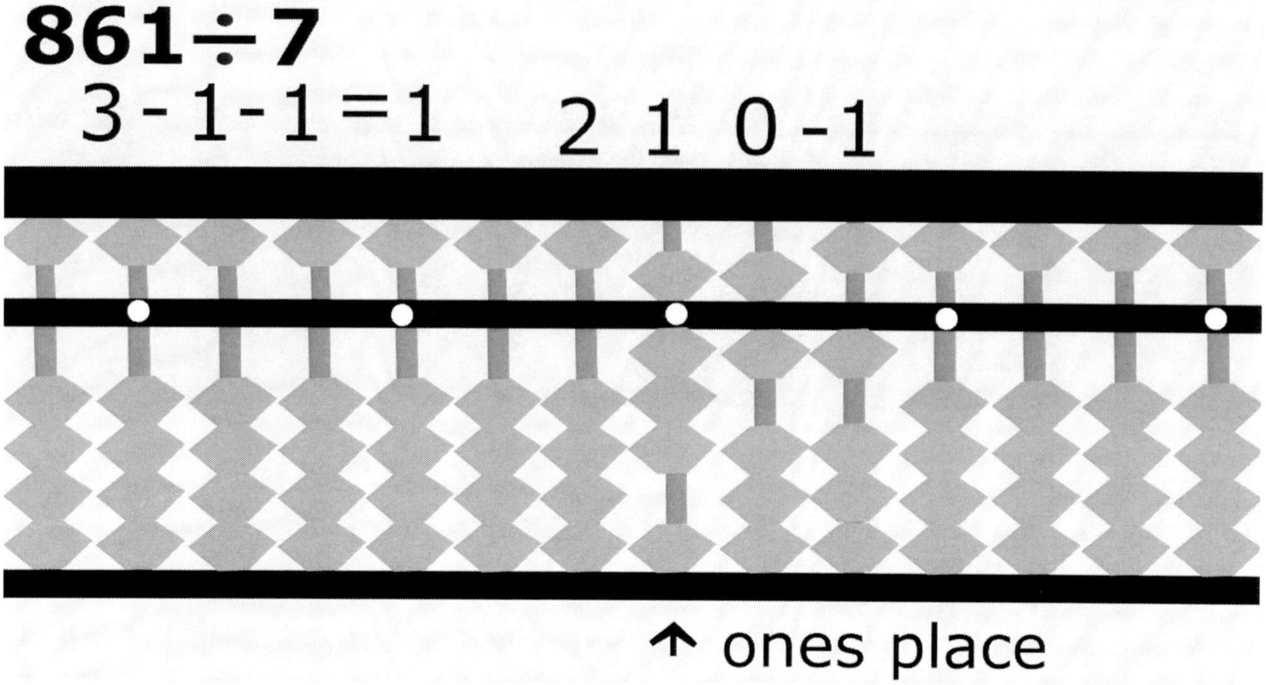

Quotient on the abacus: 8 ÷ 7 = 1 and remainder, 1 < 2, so the quotient,1, moves to the left for one place.

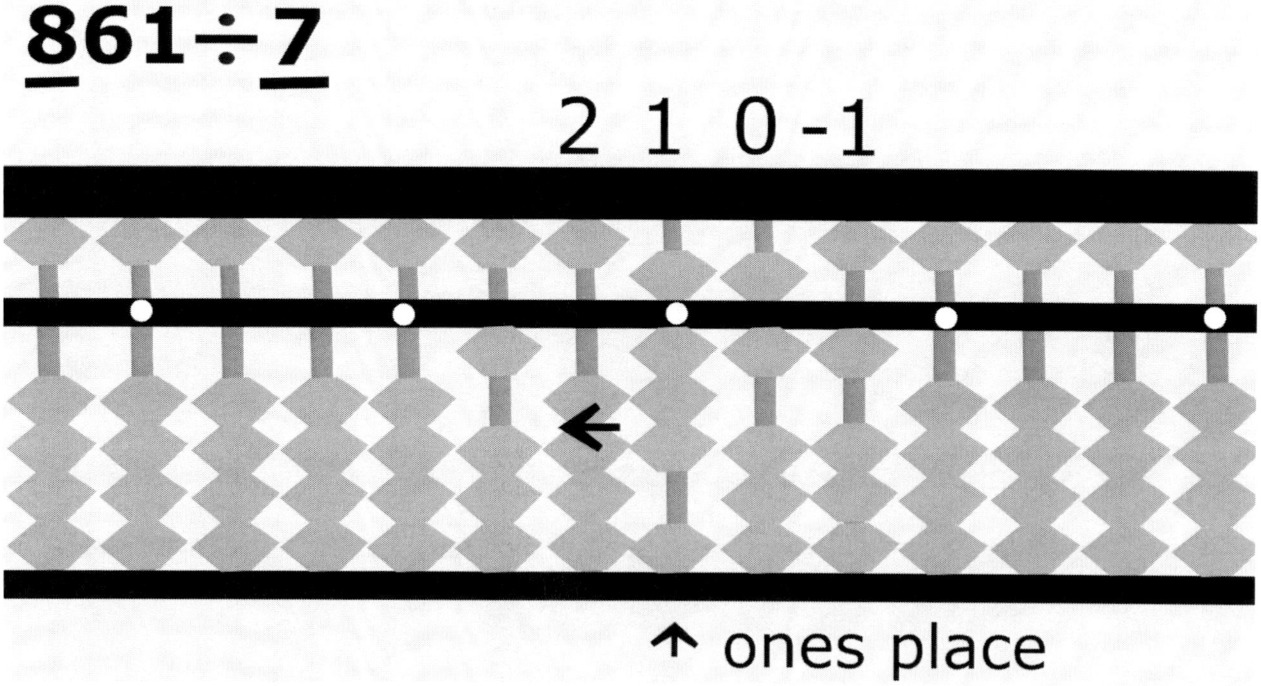

Quotient on the abacus: 8 ÷ 7 remainder is 1, the next step would be 16 ÷ 7

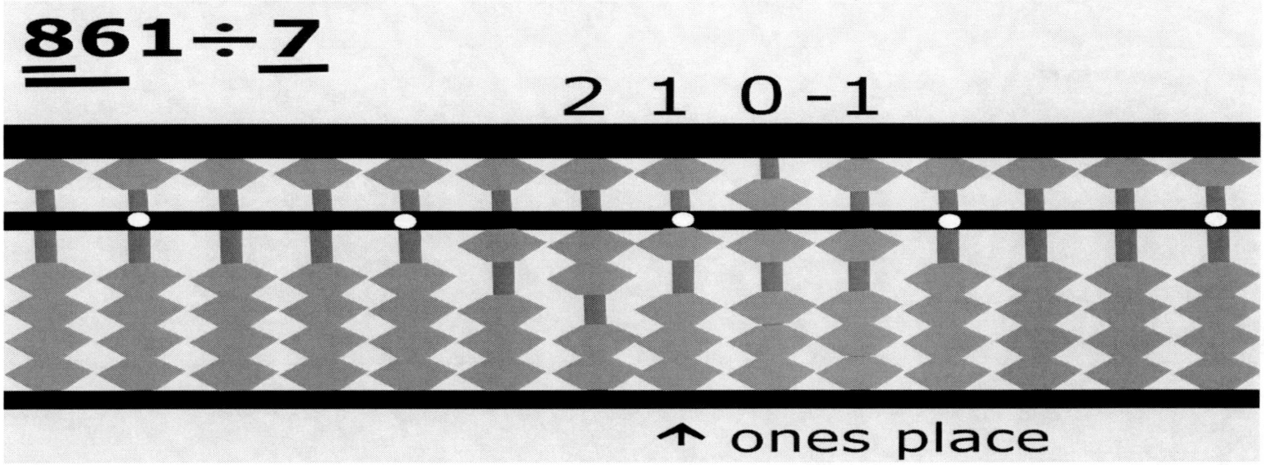

Quotient on the abacus: 16 ÷ 7 remainder is 2, the next step would be 21 ÷ 7

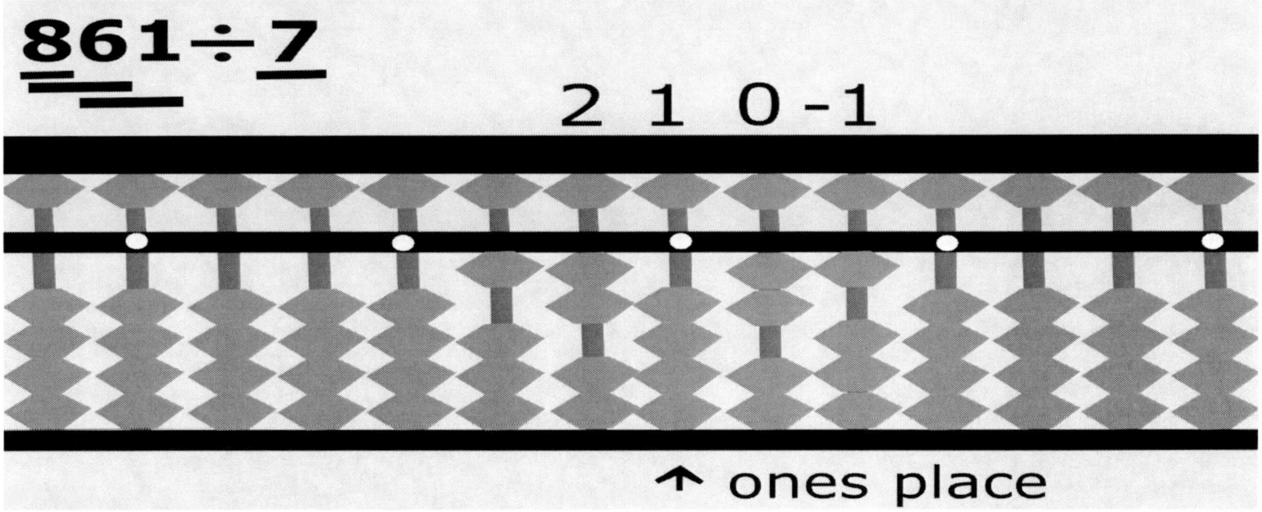

Quotient on the abacus: 861 ÷ 7 = 123

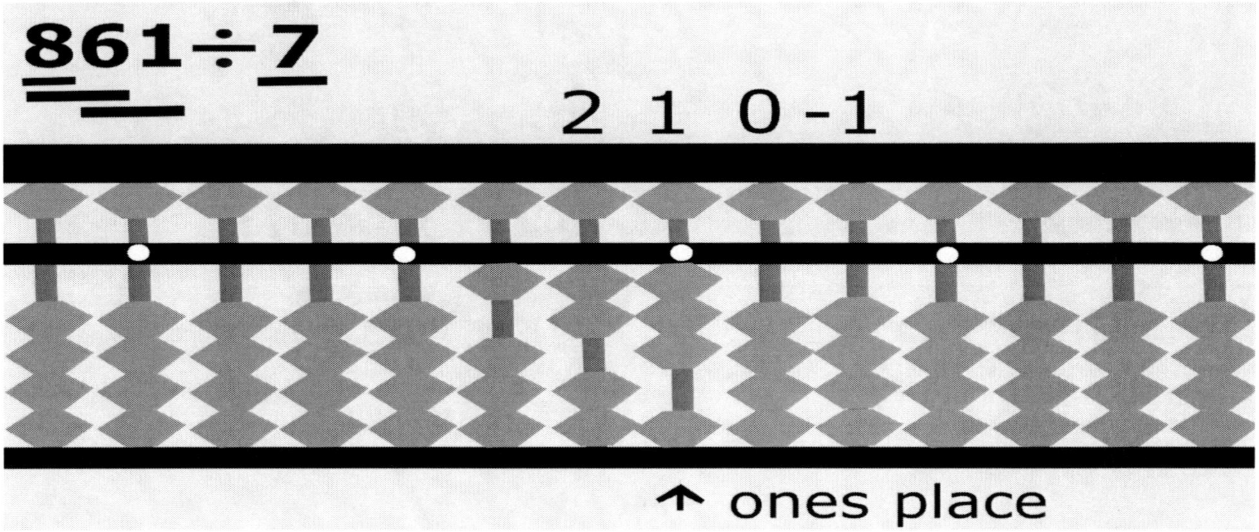

Practice for 3 digits times 1 digit

1) 954 ÷ 6 =	15) 245 ÷ 7 =
2) 480 ÷ 2 =	16) 966 ÷ 6 =
3) 852 ÷ 6 =	17) 360 ÷ 4 =
4) 322 ÷ 7 =	18) 531 ÷ 9 =
5) 524 ÷ 2 =	19) 685 ÷ 5 =
6) 216 ÷ 4 =	20) 930 ÷ 6 =
7) 342 ÷ 9 =	21) 198 ÷ 3 =
8) 736 ÷ 8 =	22) 832 ÷ 4 =
9) 444 ÷ 6 =	23) 608 ÷ 4 =
10) 153 ÷ 9 =	24) 185 ÷ 5 =
11) 256 ÷ 8 =	25) 555 ÷ 3 =
12) 138 ÷ 6 =	26) 894 ÷ 6 =
13) 567 ÷ 3 =	27) 684 ÷ 6 =
14) 180 ÷ 5 =	28) 288 ÷ 8 =

Albert Einstein:

As far as the laws of mathematics refer to reality, they are not certain, and as far as they are certain, they do not refer to reality.

9. three digits divided by two digits

Dividend on the abacus: 3 digits of dividend – 1 division symbol – 2 digits of divisor = dividend position on the abacus. The abacus position for 945 is 0, on the tenth place.

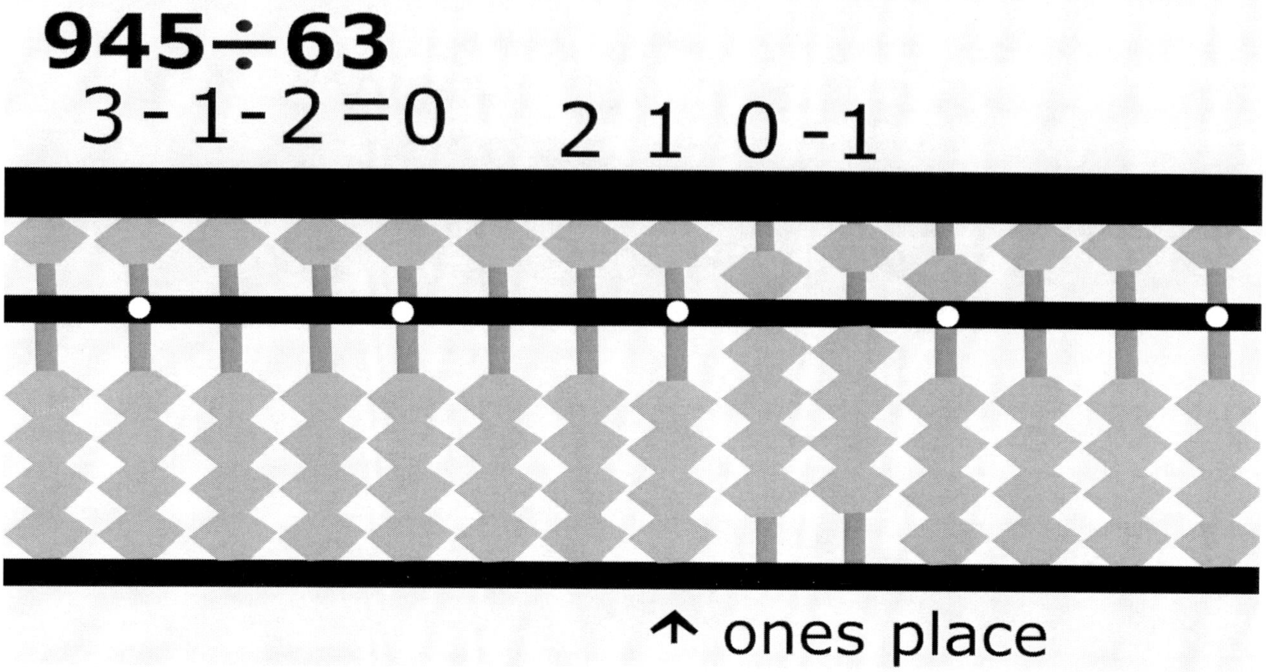

Quotient on the abacus: 94 ÷ 63 = 1 and remainder, 1 < 9, so the quotient,1, moves to the left for one place.

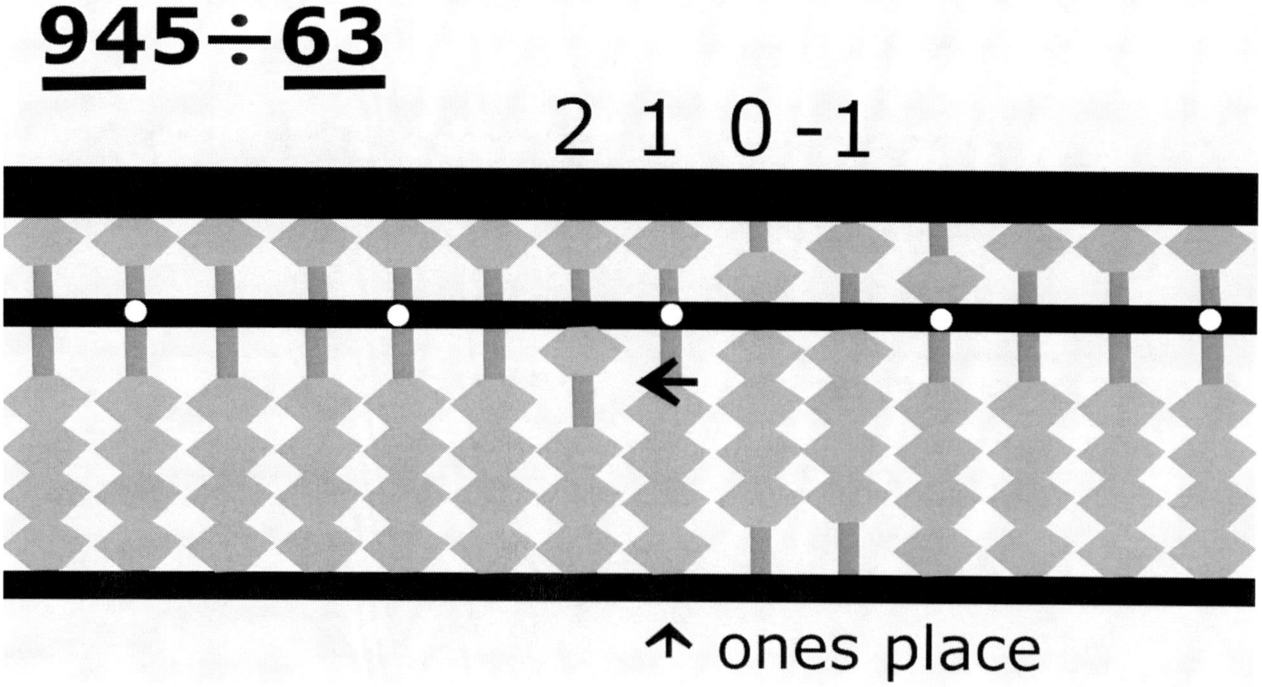

Quotient on the abacus: 94 ÷ 67 remainder is 31, the next step would be 315 ÷ 63

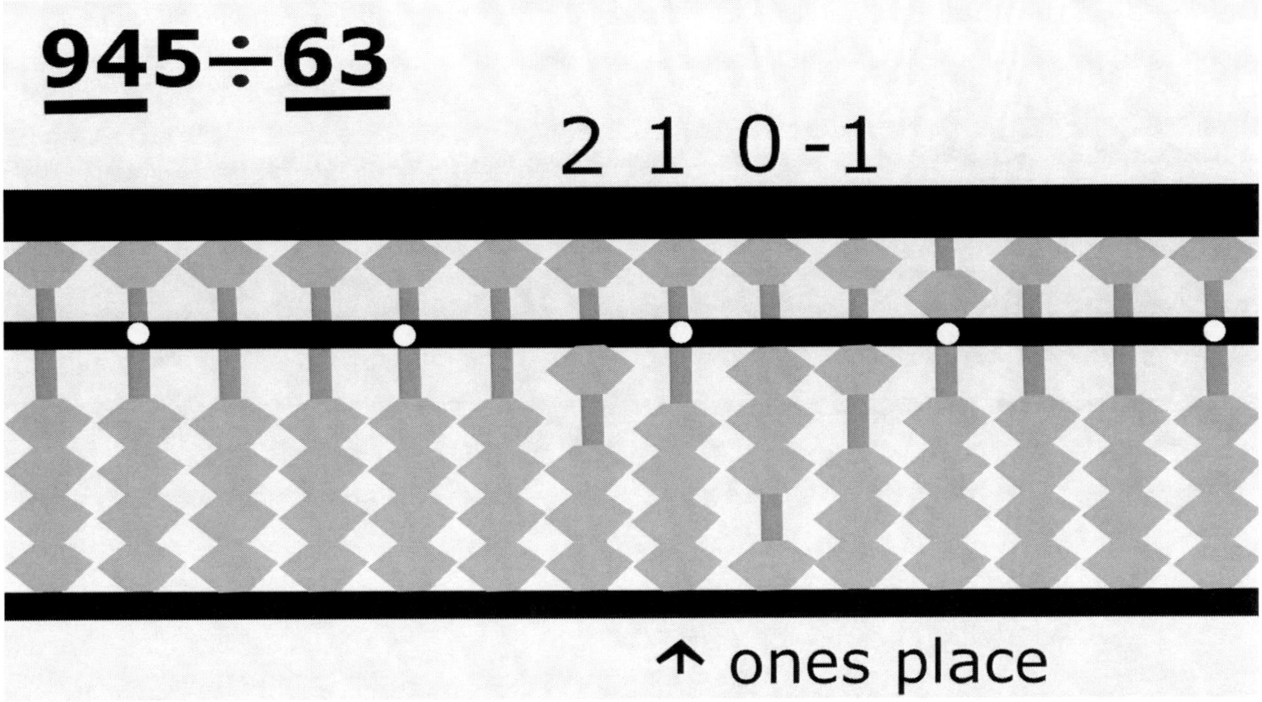

Quotient on the abacus: 945 ÷ 63 = 15

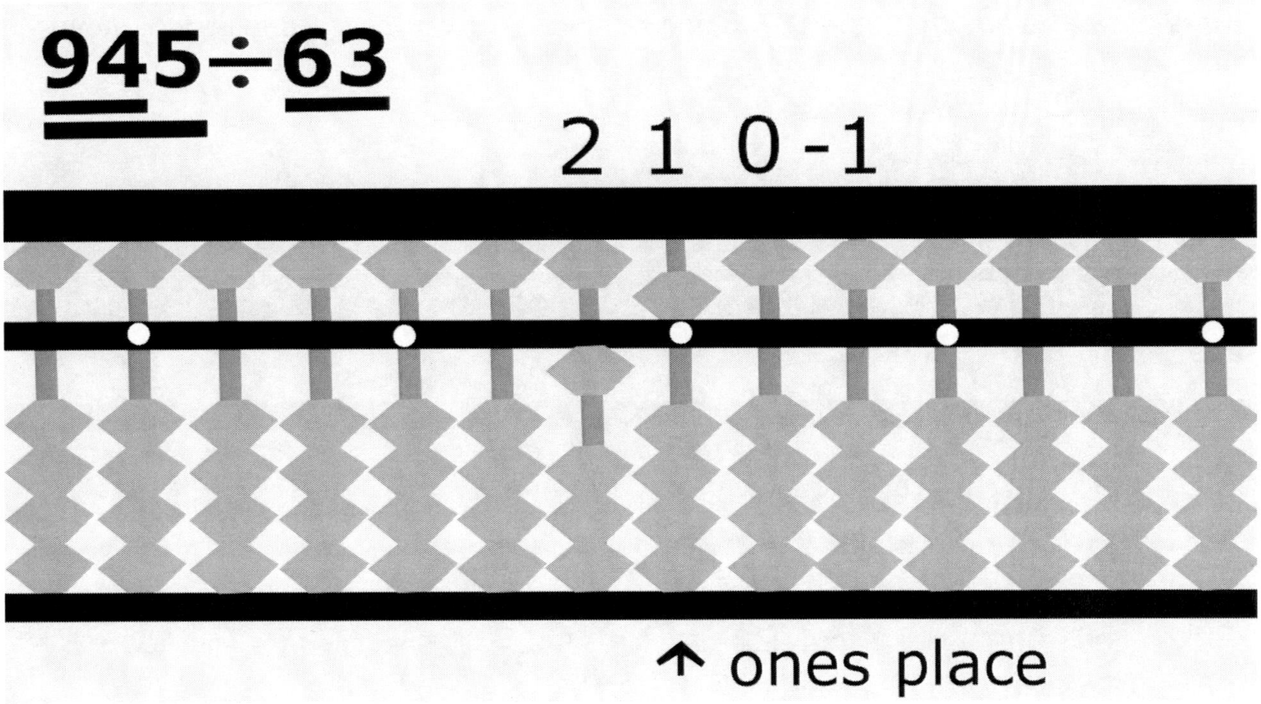

Dividend on the abacus: 3 digits of dividend – 1 division symbol – 2 digits of divisor = dividend position on the abacus. The abacus position for 558 is 0, on the tenth place.

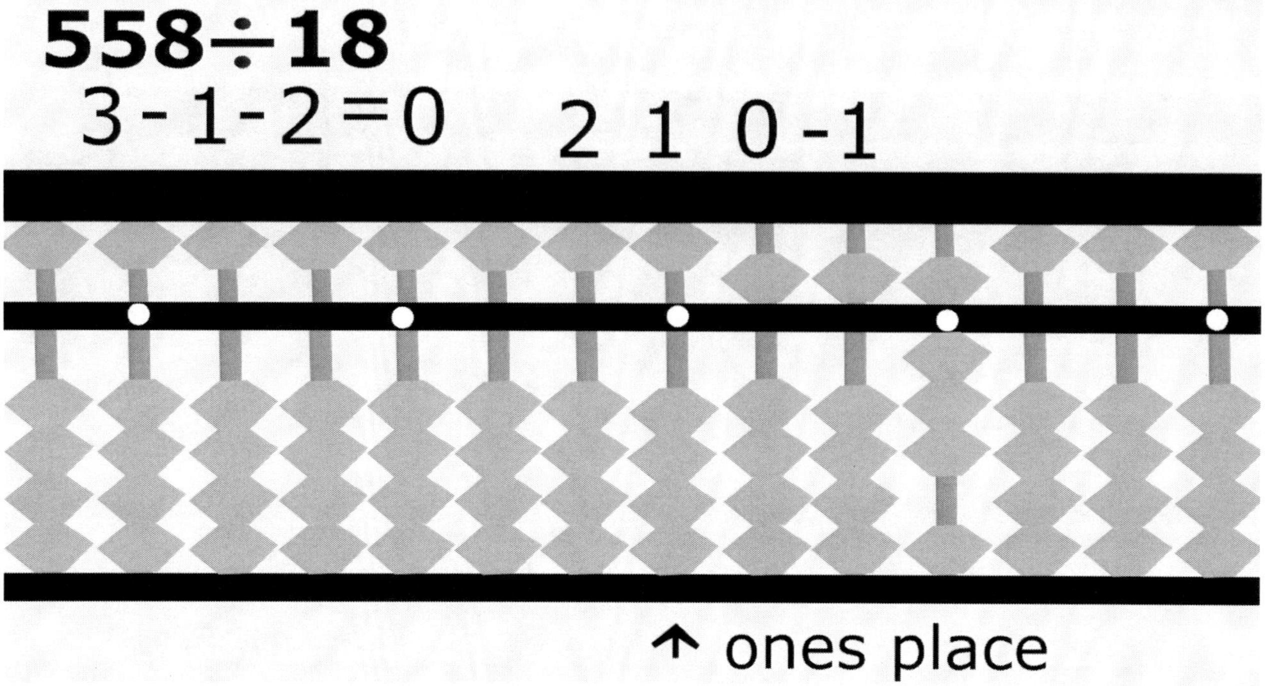

Quotient on the abacus: 55 ÷ 18 = 3 and remainder, 3 < 5, so the quotient, 3, moves to the left for one place.

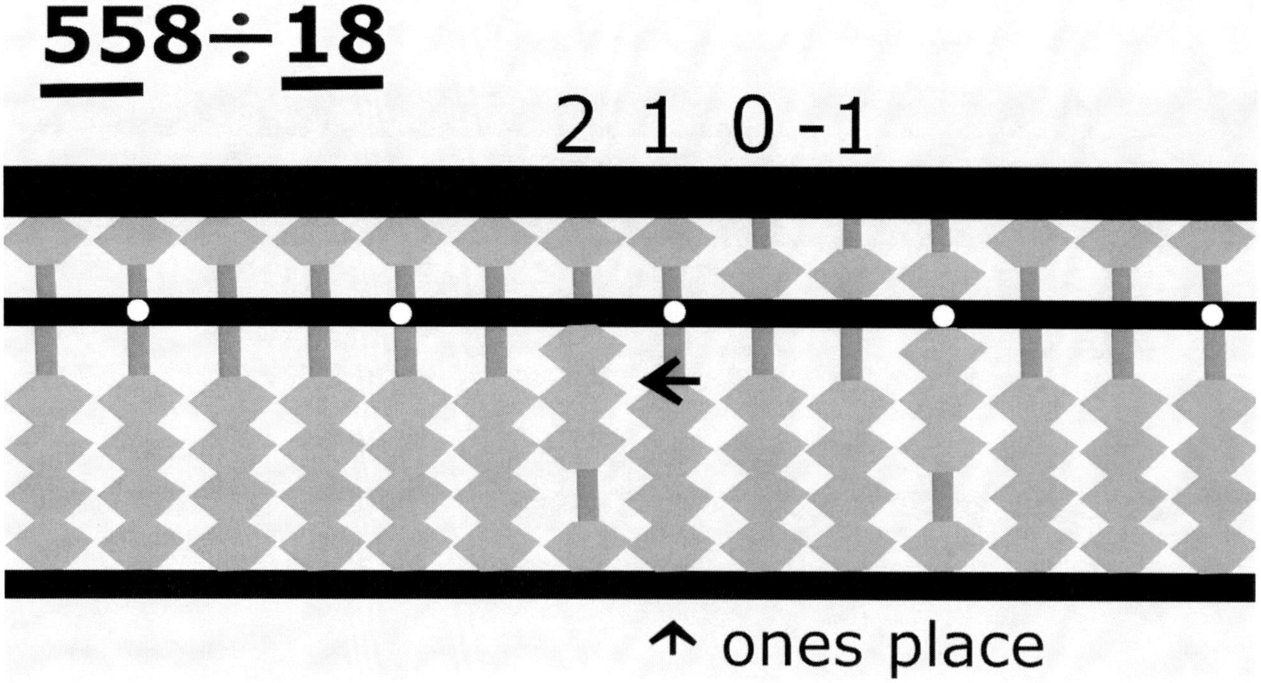

Quotient on the abacus: 55 ÷ 18 remainder is 01, the next step would be 018 ÷ 18

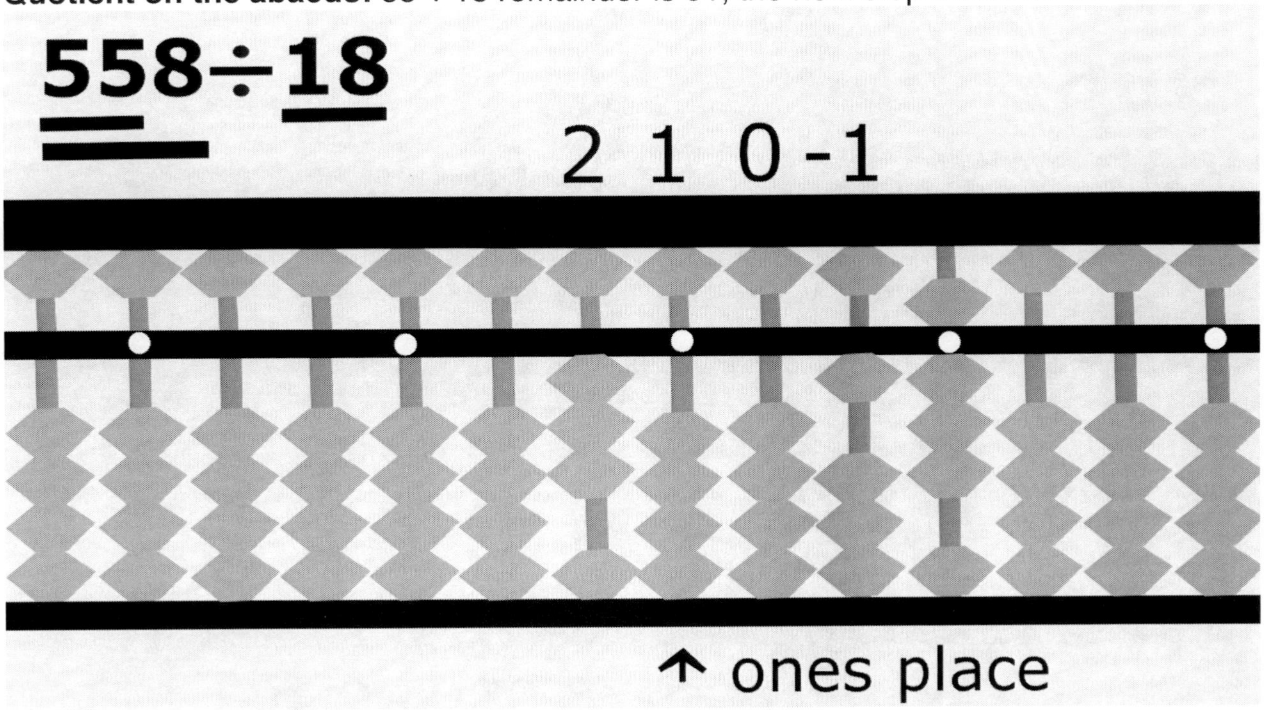

Quotient on the abacus: 558 ÷ 18 = 31

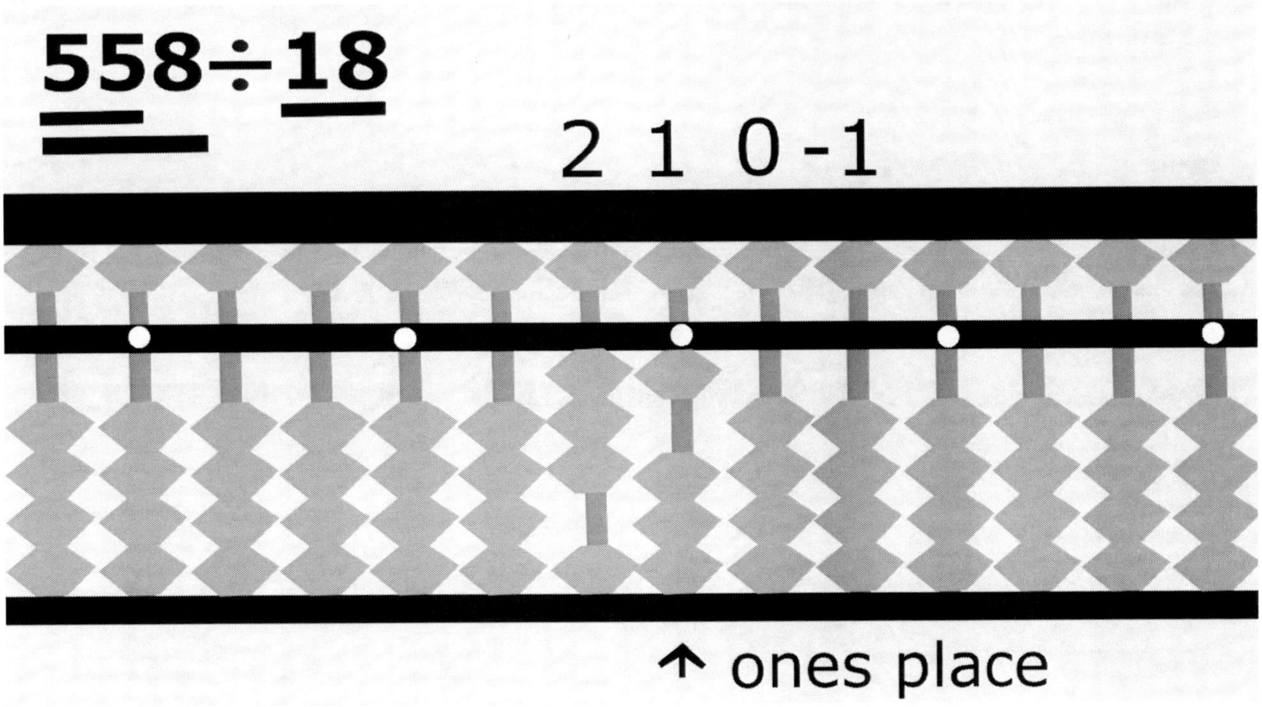

Dividend on the abacus: 3 digits of dividend – 1 division symbol – 2 digits of divisor = dividend position on the abacus. The abacus position for 966 is 0, on the tenth place.

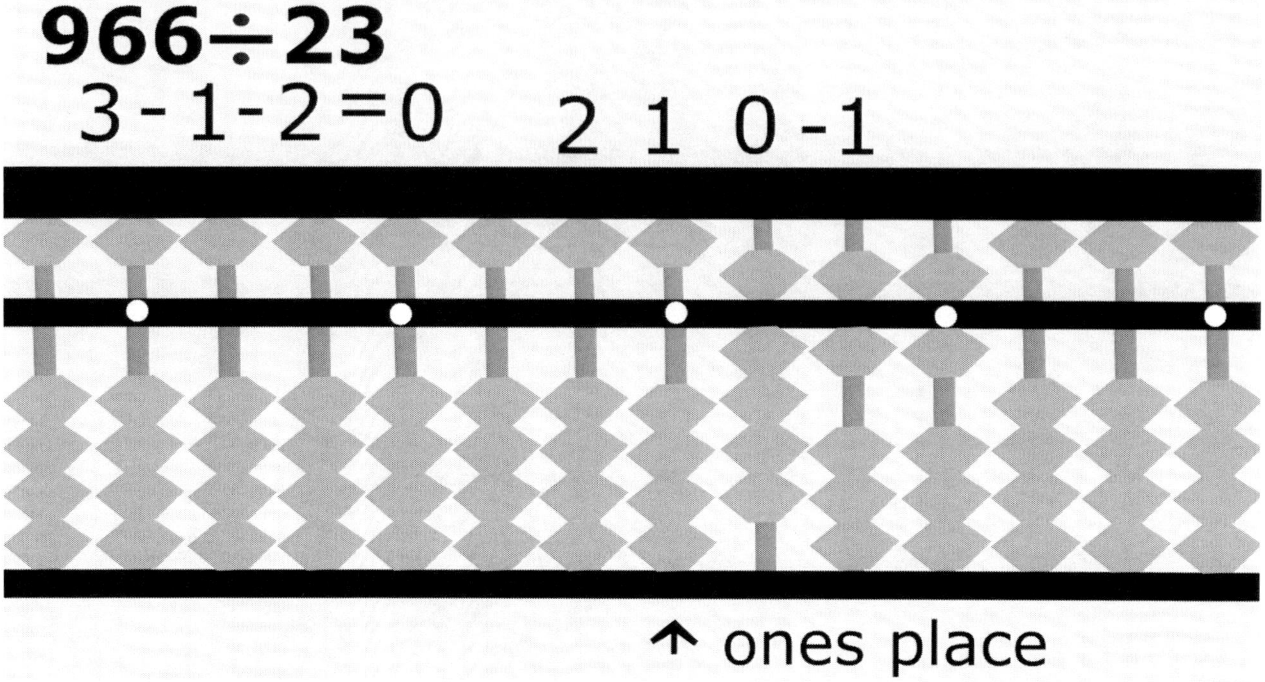

Quotient on the abacus: 96 ÷ 23 = 4 and remainder, 4 < 9, so the quotient, 4, moves to the left for one place.

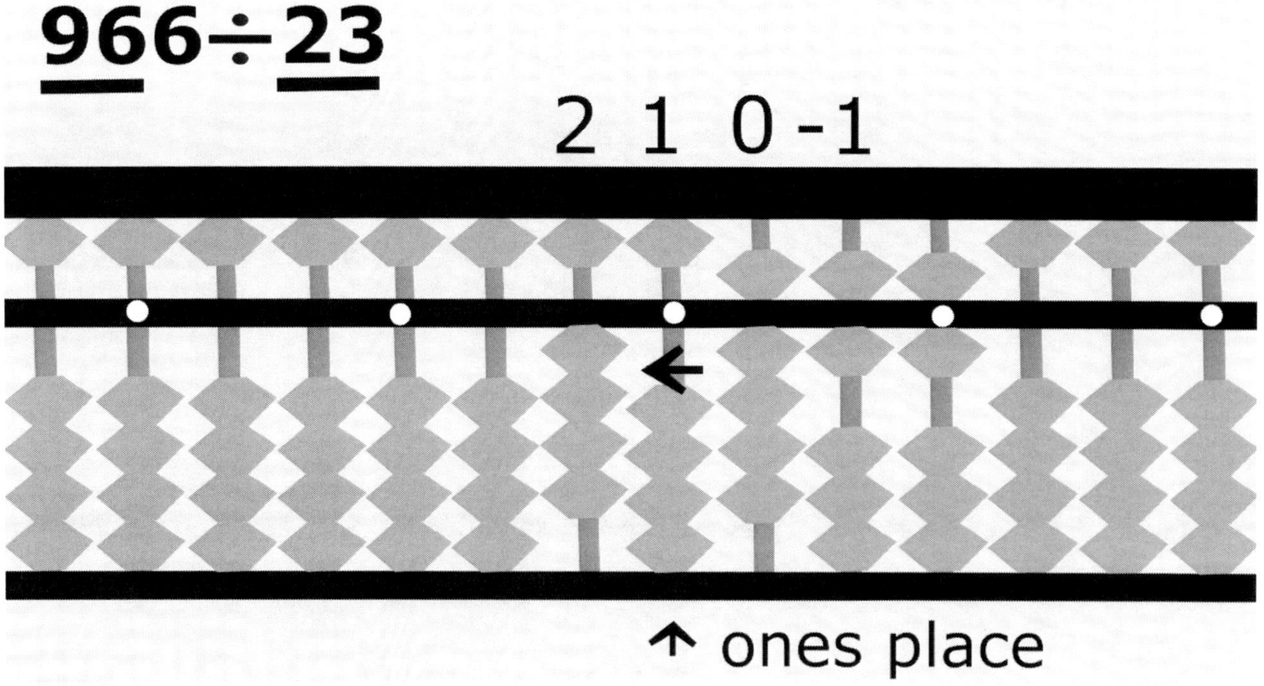

Quotient on the abacus: 96 ÷ 23 remainder is 04, the next step would be 046 ÷ 23

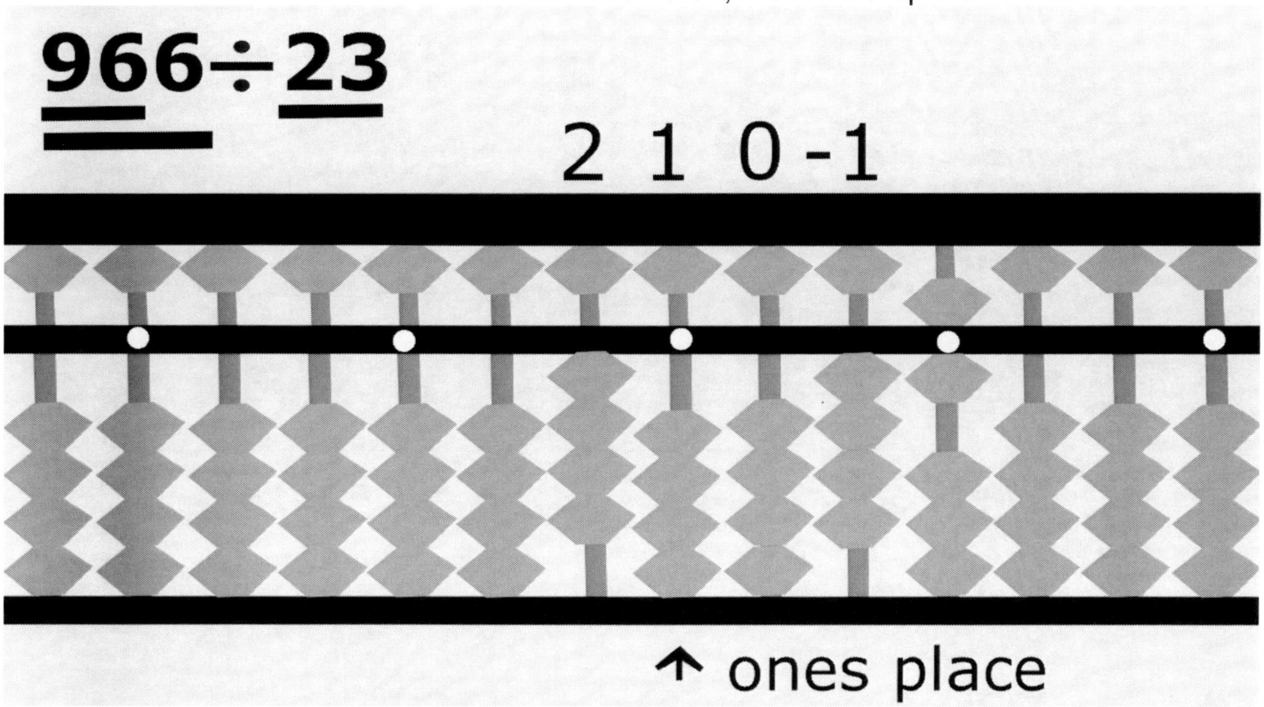

Quotient on the abacus: 966 ÷ 23 = 42

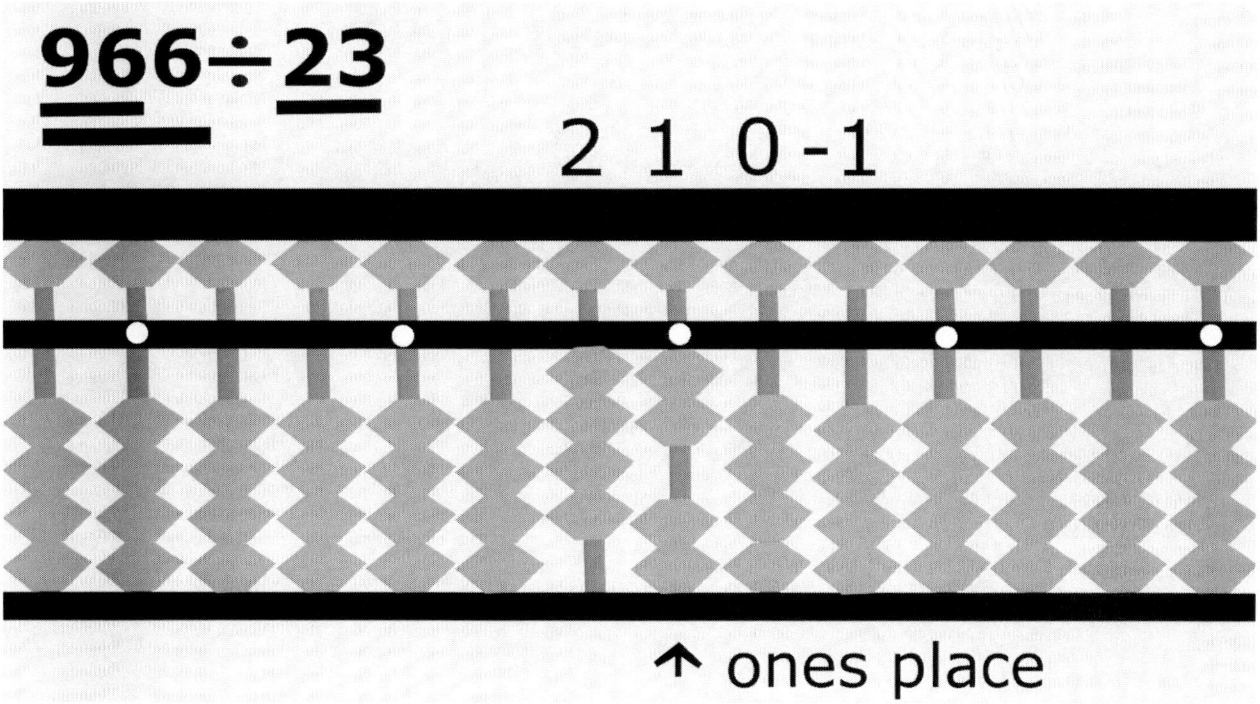

Practice for 3 digits divided by 2 digits

1) 242 ÷ 22 =	15) 874 ÷ 38 =
2) 495 ÷ 99 =	16) 518 ÷ 74 =
3) 990 ÷ 55 =	17) 225 ÷ 25 =
4) 833 ÷ 49 =	18) 552 ÷ 46 =
5) 494 ÷ 38 =	19) 800 ÷ 32 =
6) 686 ÷ 98 =	20) 490 ÷ 35 =
7) 930 ÷ 62 =	21) 884 ÷ 26 =
8) 378 ÷ 27 =	22) 280 ÷ 28 =
9) 644 ÷ 28 =	23) 845 ÷ 65 =
10) 949 ÷ 73 =	24) 679 ÷ 97 =
11) 896 ÷ 64 =	25) 780 ÷ 60 =
12) 885 ÷ 59 =	26) 871 ÷ 67 =
13) 392 ÷ 98 =	27) 935 ÷ 55 =
14) 819 ÷ 63 =	28) 806 ÷ 31 =

Albert Einstein:

Do not worry about your difficulties in mathematics. I can assure you mine are still greater

10. Division with decimal

The division algorithm is the same for numbers containing decimals. However, only integers' digits are used to identify the dividend's position on the abacus. Let's use the examples previously introduced for illustrations.

The integers for 4.8 ÷ 2 are 4 and 2. The position of 4.8 would be 1 – 1 (division symbol) – 1= -1

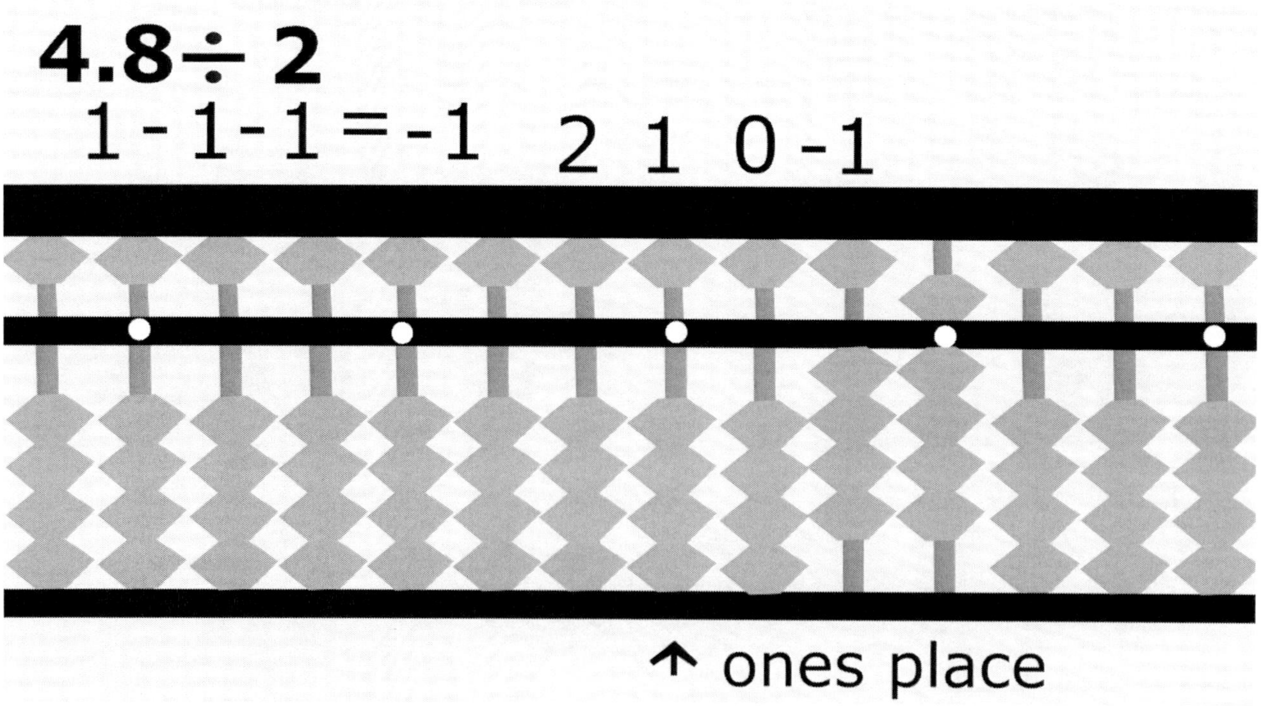

The integers for 12.3 ÷ 3 are 12 and 3. The position of 12.3 would be 2 – 1 – 1= 0

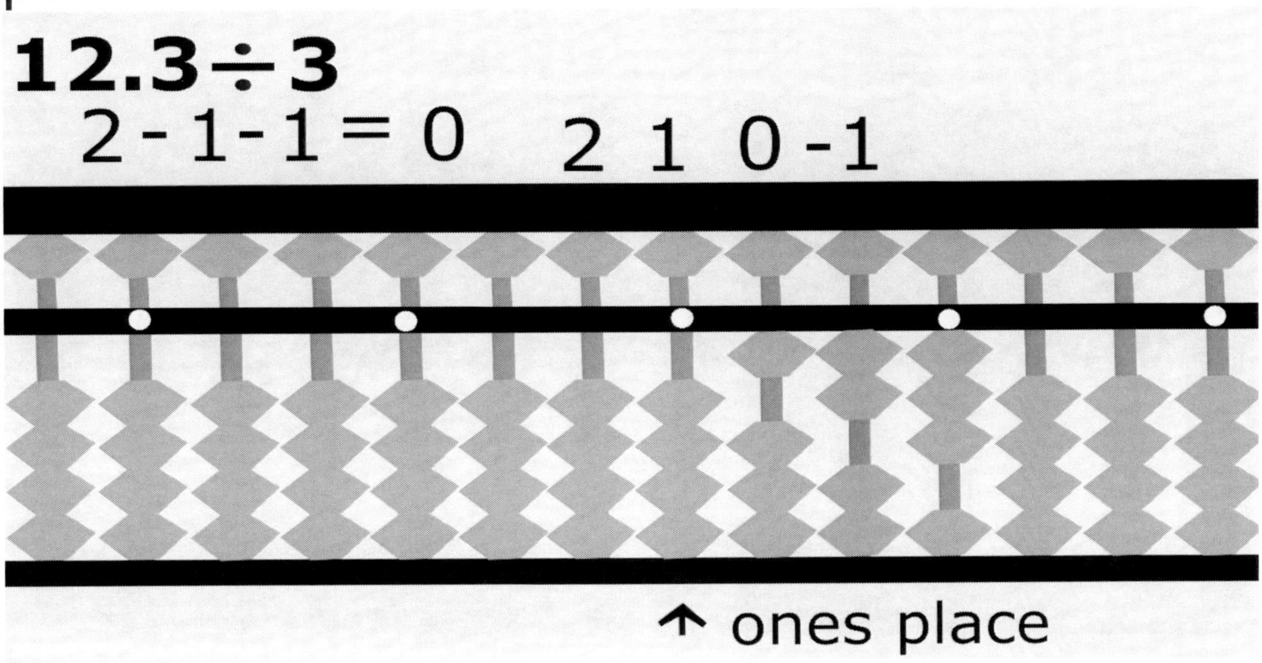

The integers for 94.5 ÷ 63 are 94 and 63. The position of 94.5 would be 2 – 1 – 2 = -1

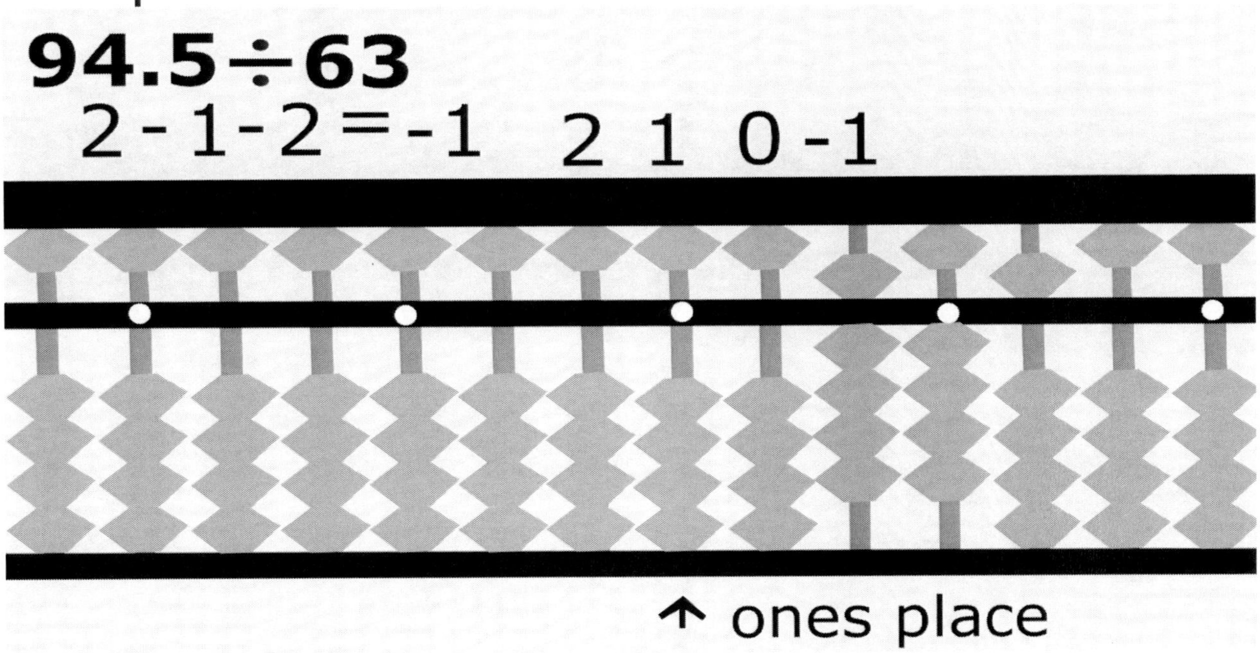

Dividend on the abacus: 0 digits of dividend – 1 division symbol – 1 digits of divisor = dividend position on the abacus. The abacus position for 0.95 is -2, on the thousandth place.

0.95 ÷ 5

0 - 1 - 1 = -2

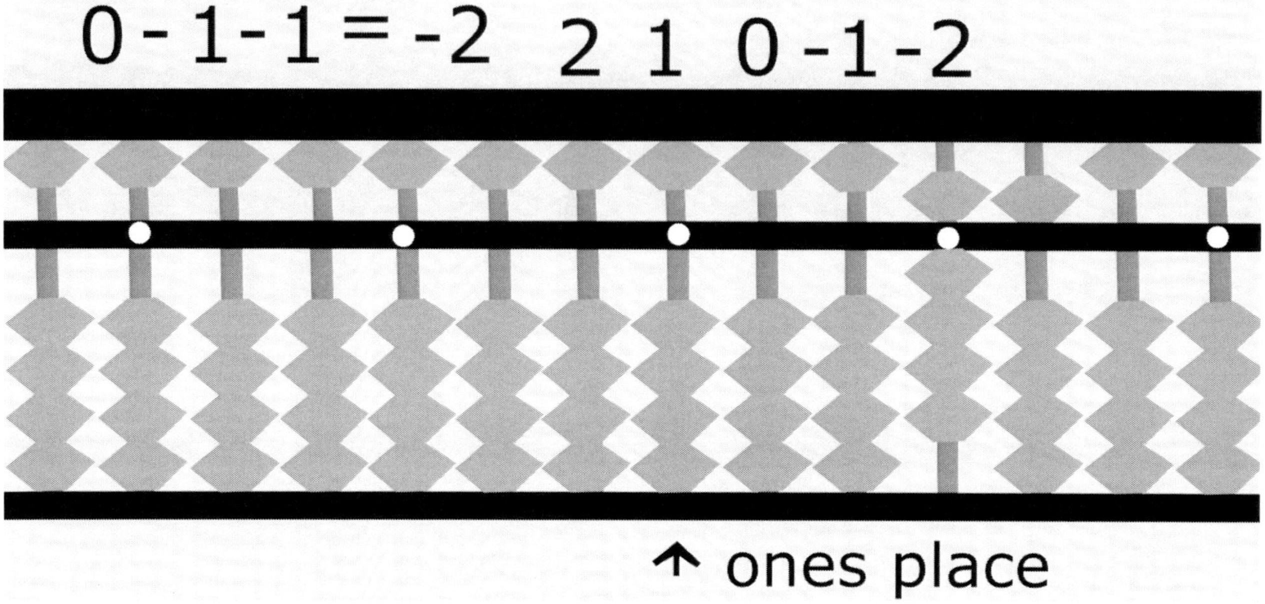

Quotient on the abacus: 9 ÷ 5 = 1 and remainder, 1 < 9, so the quotient, 1, moves to the left for one place.

0.95 ÷ 5

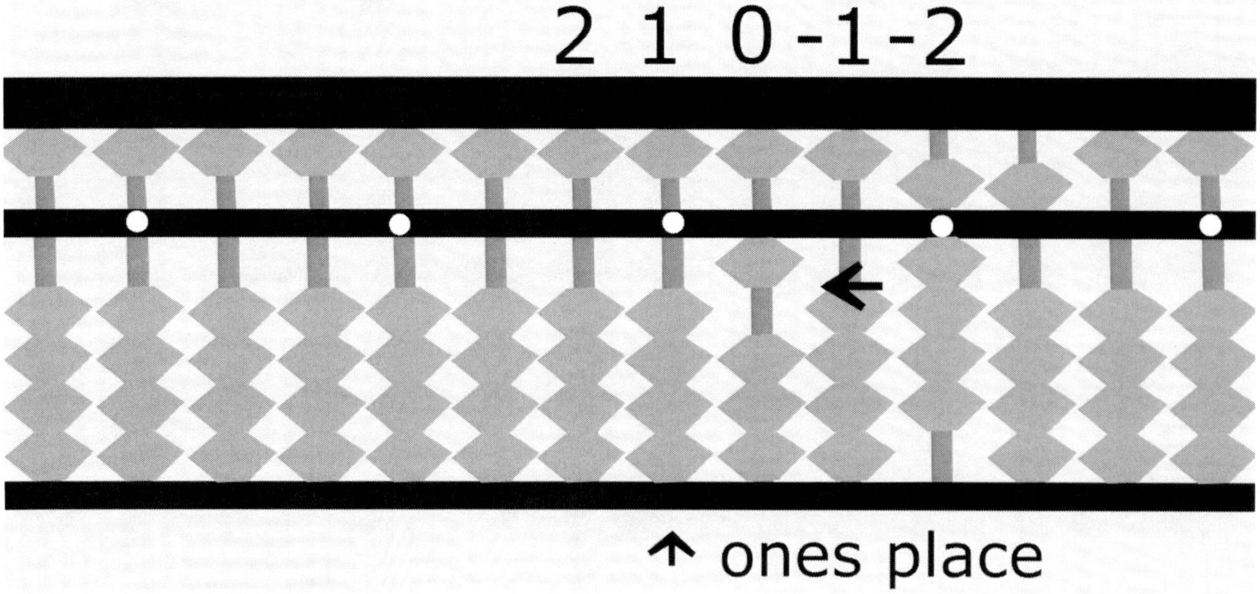

Quotient on the abacus: 9 ÷ 5 remainder is 4, the next step would be 45 ÷ 5

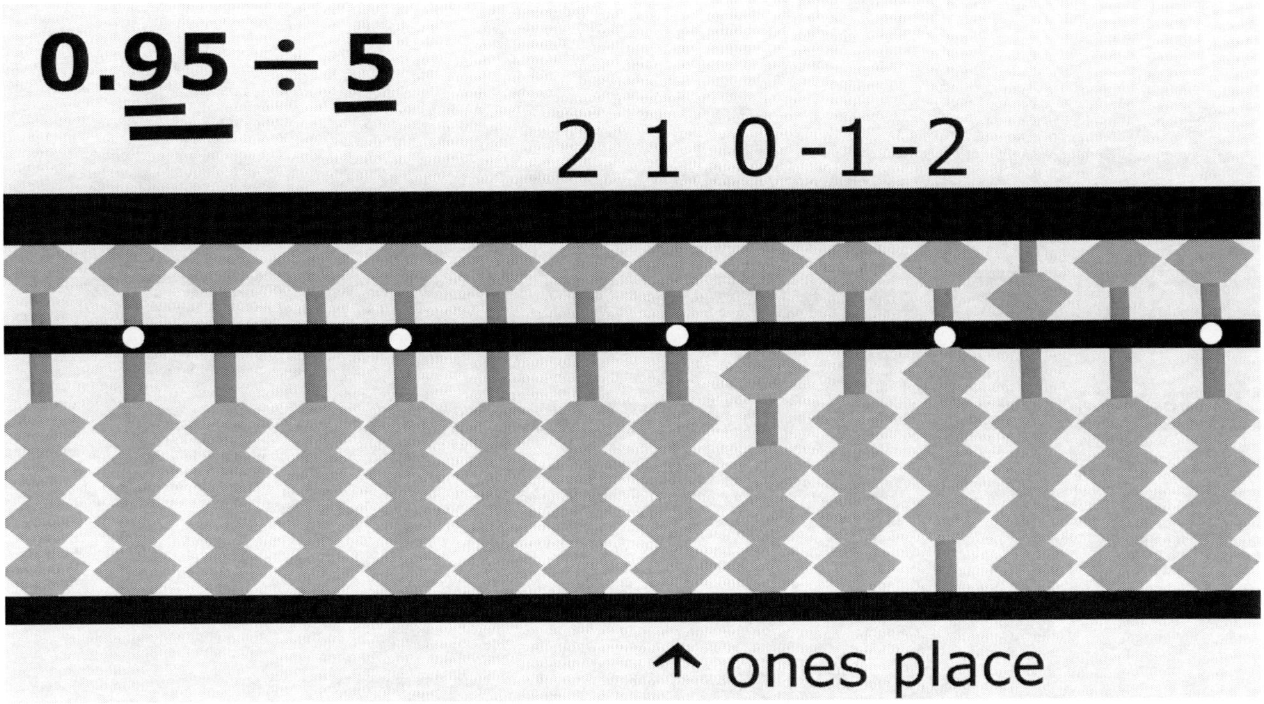

Quotient on the abacus: 0.95 ÷ 5 = 0.19

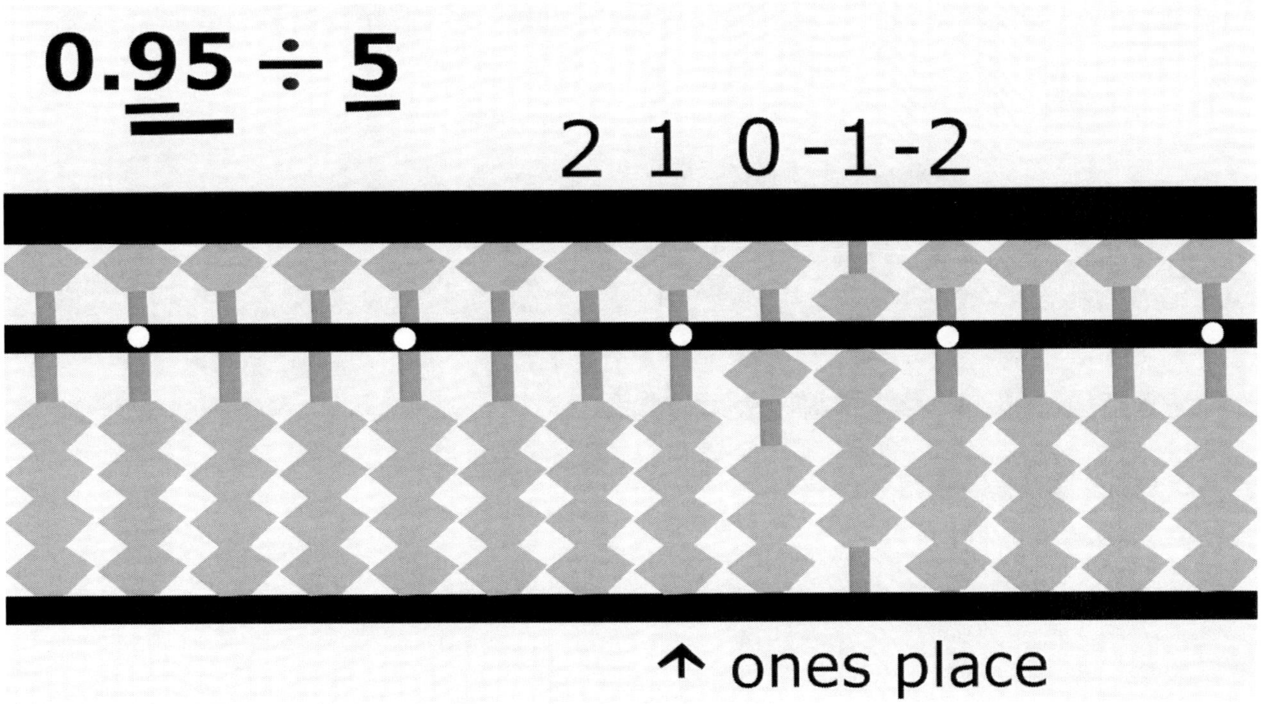

Dividend on the abacus: 0 digits of dividend – 1 division symbol – 1 digits of divisor = dividend position on the abacus. The abacus position for 2.58 is -1, on the hundredth place.

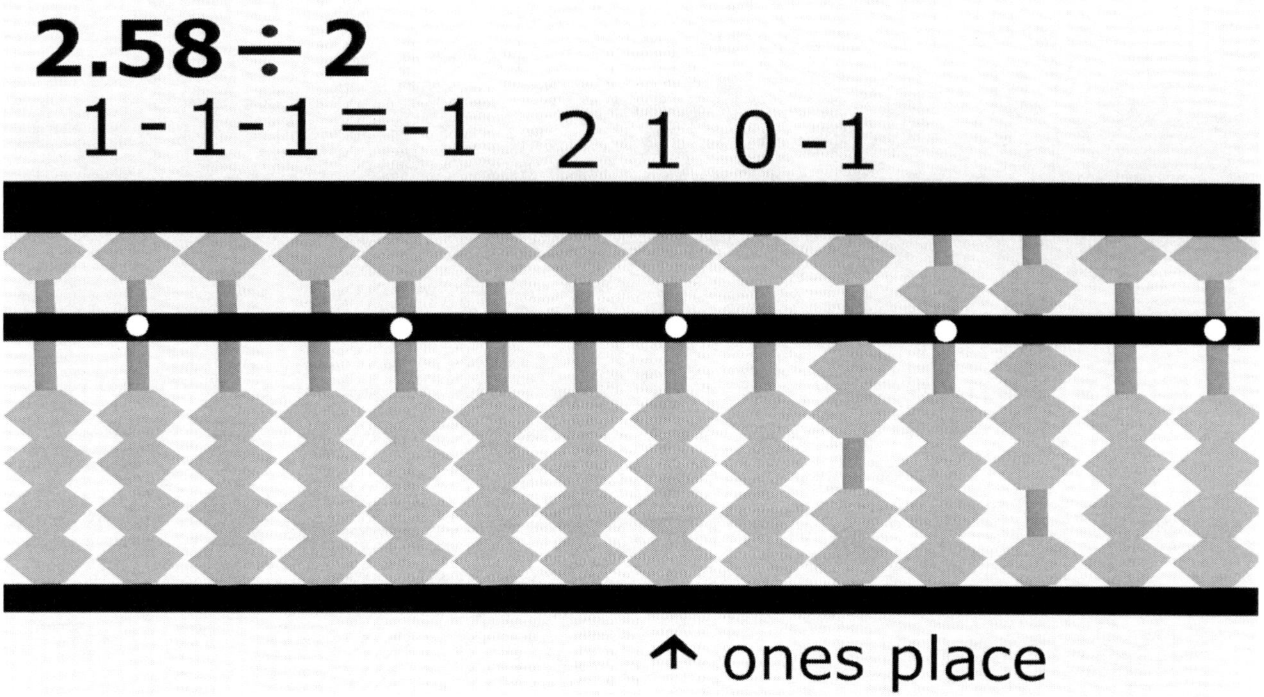

Quotient on the abacus: 2 ÷ 2 = 1, 1 < 9, so the quotient, 1, moves to the left for one place.

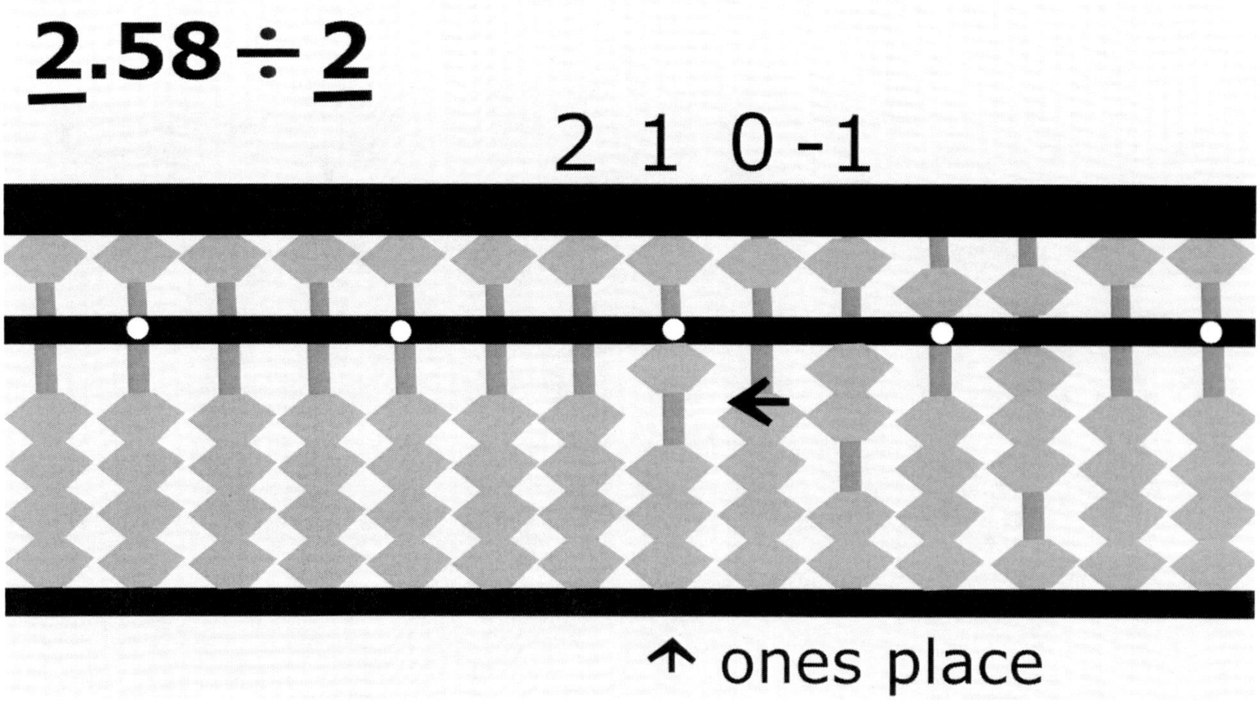

Quotient on the abacus: 2 ÷ 2 remainder is 0, the next step would be 05 ÷ 2

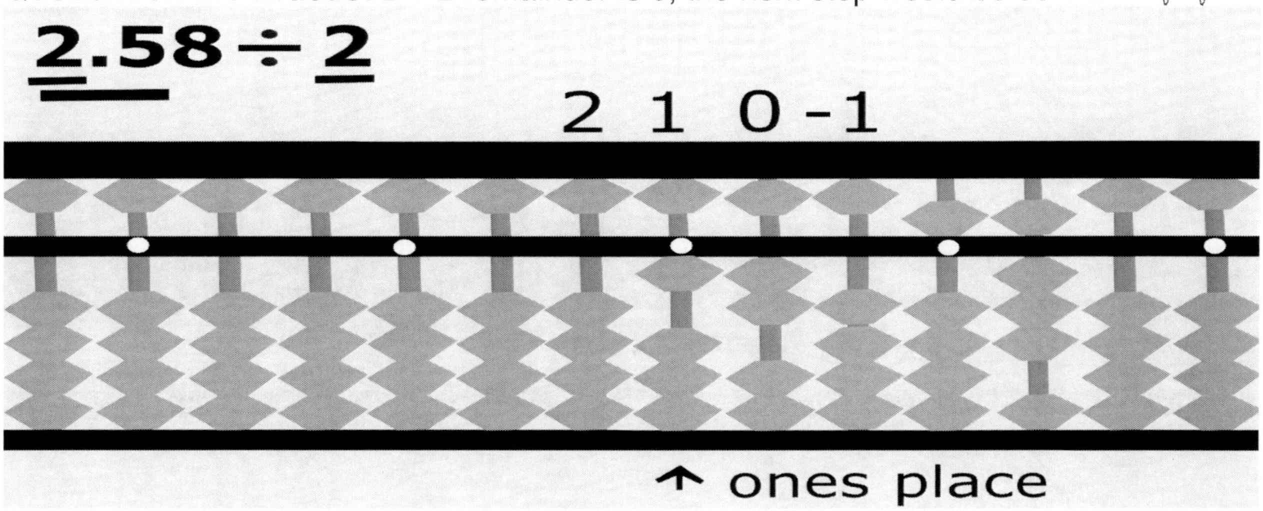

Quotient on the abacus: 05 ÷ 2 remainder is 1, the next step would be 18 ÷ 2

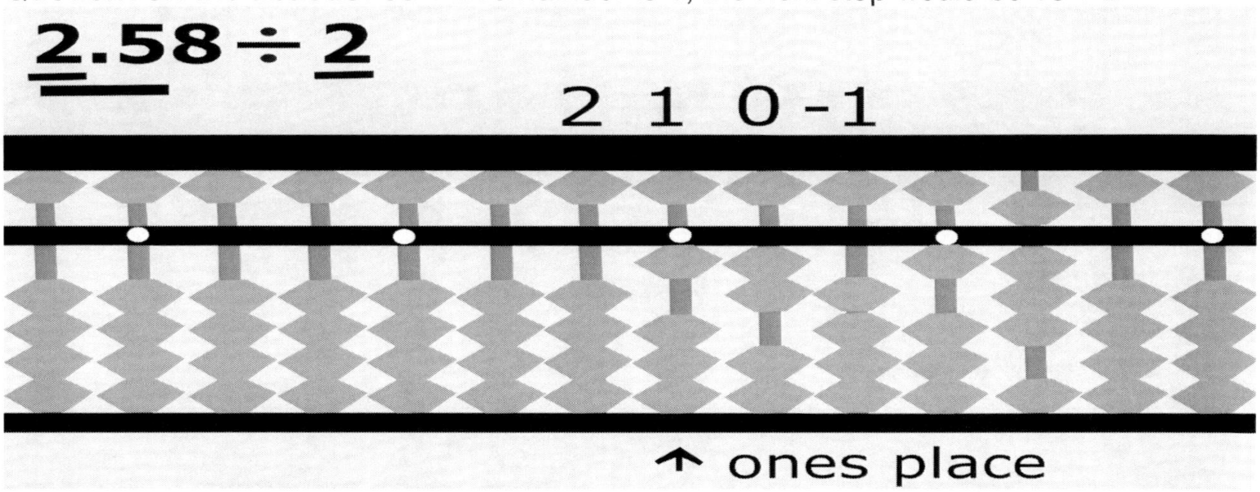

Quotient on the abacus: 2.58 ÷ 2 = 1.29

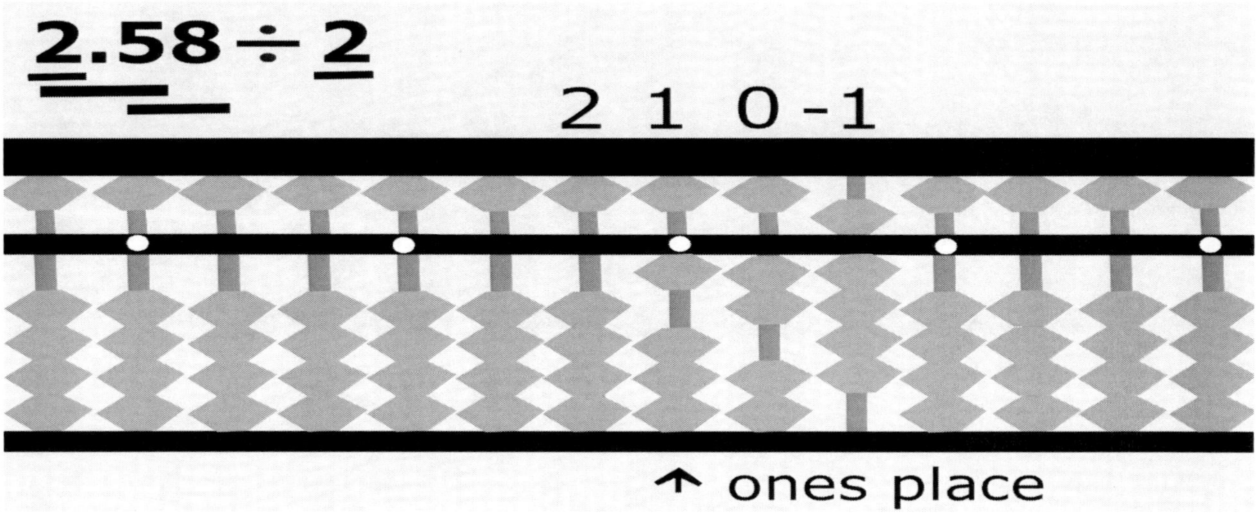

Dividend on the abacus: 1 digit of dividend – 1 division symbol – 2 digits of divisor = dividend position on the abacus. The abacus position for 9.66 is -2, on the thousandth place.

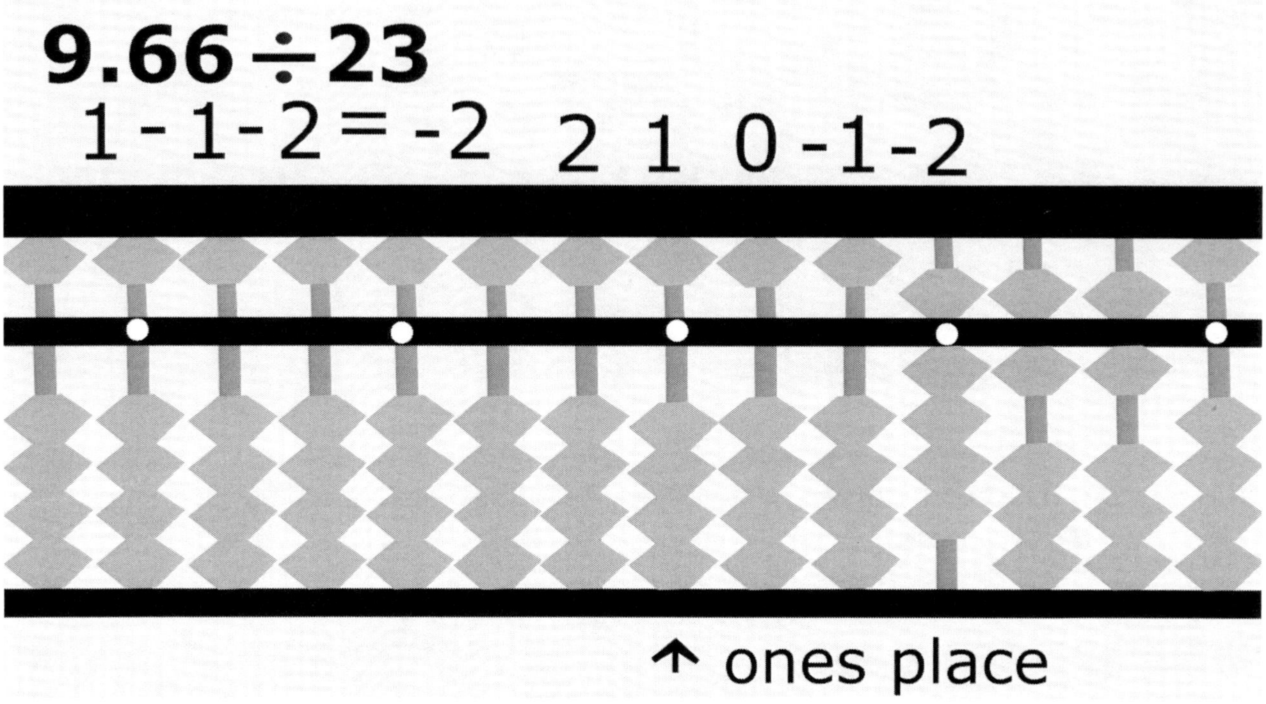

Quotient on the abacus: 96÷ 23 = 4 and remainder, 4 < 9, so the quotient, 4, moves to the left for one place.

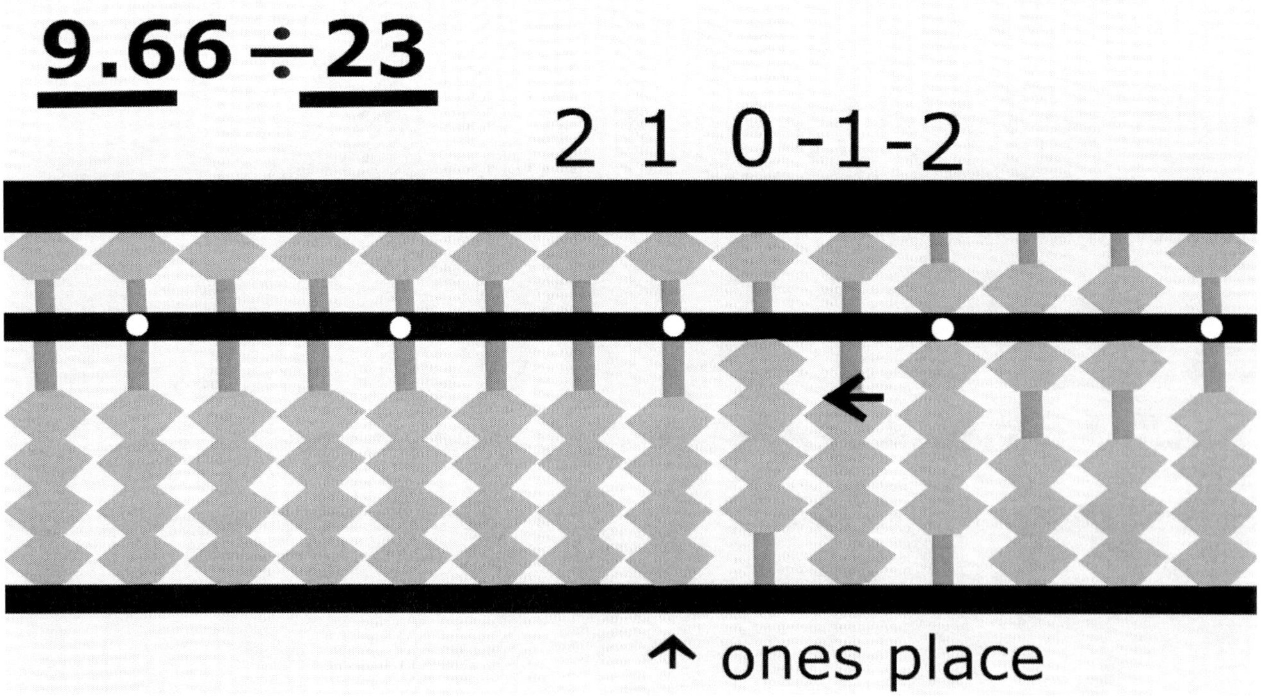

Quotient on the abacus: 96 ÷ 23 remainder is 4, the next step would be 46 ÷ 23

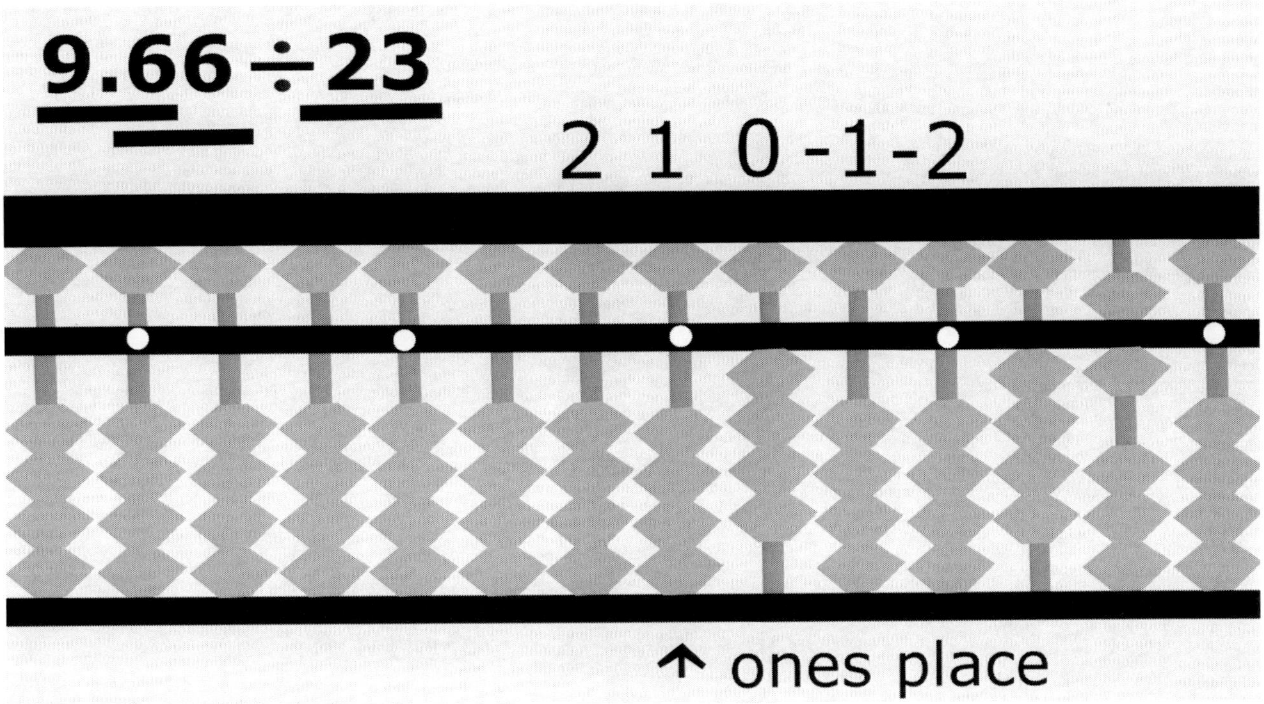

Quotient on the abacus: 9.66 ÷ 23 = 0.42

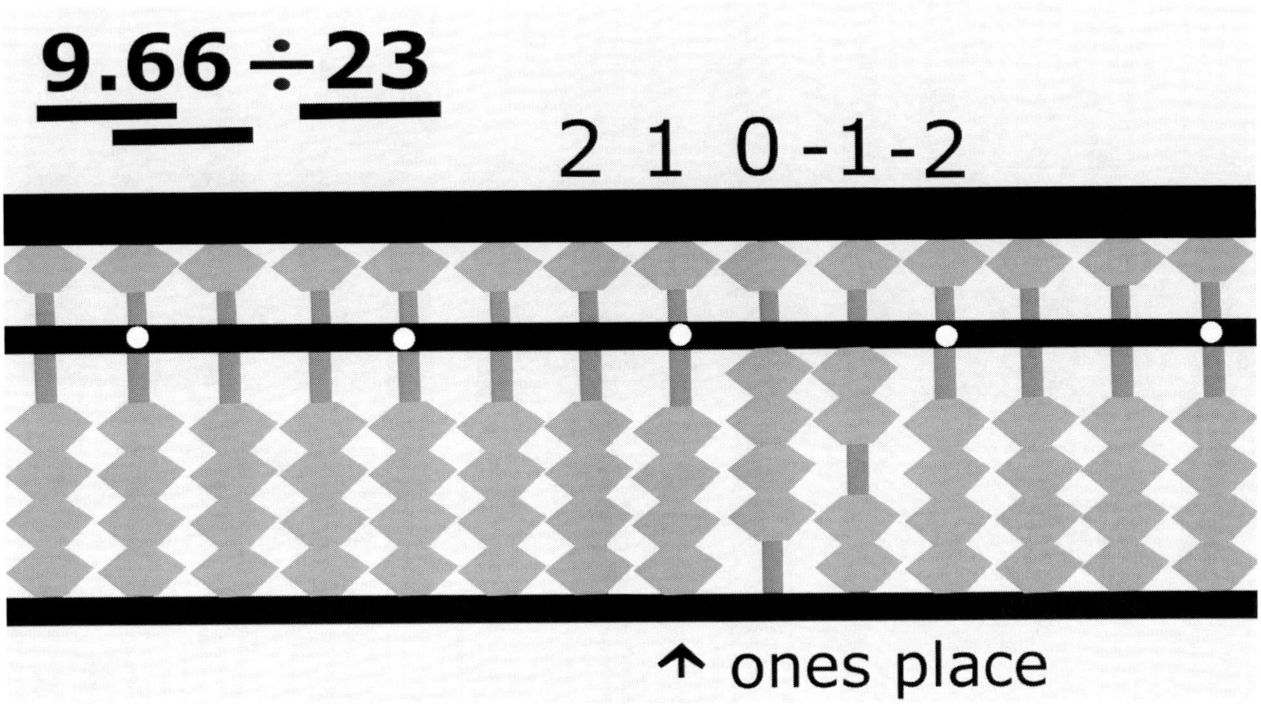

Practice for decimal division

1) 81 ÷ 2.7 =	15) 83.6 ÷ 3.8 =
2) 89.64 ÷ 99.6 =	16) 57.6 ÷ 4.8 =
3) 70.2 ÷ 5.4 =	17) 65.57 ÷ 8.3 =
4) 44.1 ÷ 4.9 =	18) 34.5 ÷ 4.6 =
5) 123 ÷ 8.2 =	19) 72.5 ÷ 2.9 =
6) 323.4 ÷ 9.8 =	20) 99.4 ÷ 3.5 =
7) 5.22 ÷ 8.7 =	21) 156 ÷ 6.5 =
8) 40.5 ÷ 2.7 =	22) 282.2 ÷ 8.3 =
9) 22.5 ÷ 2.5 =	23) 55.18 ÷ 8.9 =
10) 9.75 ÷ 6.5 =	24) 6.24 ÷ 4.8 =
11) 8.33 ÷ 4.9 =	25) 77.6 ÷ 9.7 =
12) 141.6 ÷ 5.9 =	26) 52.2 ÷ 8.7 =
13) 26.541 ÷ 98.3 =	27) 94.9 ÷ 7.3 =
14) 825.5 ÷ 63.5 =	28) 20.14 ÷ 3.8 =

Yay !!!
You DID it .

Answer Key

p. 30 (2 digits divided by 1 digit)

1	2	3	4	5	6	7	8	9	10
11	**16**	**13**	**27**	**23**	**18**	**13**	**25**	**17**	**19**
11	12	13	14	15	16	17	18	19	20
44	**13**	**28**	**15**	**13**	**29**	**17**	**11**	**23**	**47**
21	22	23	24	25	26	27	28		
14	**26**	**10**	**11**	**10**	**12**	**24**	**22**		

p. 37 (3 digits divided by 1 digit)

1	2	3	4	5	6	7	8	9	10
159	**240**	**142**	**46**	**262**	**54**	**38**	**92**	**74**	**17**
11	12	13	14	15	16	17	18	19	20
32	**23**	**189**	**36**	**35**	**161**	**90**	**59**	**137**	**155**
21	22	23	24	25	26	27	28		
66	**208**	**152**	**37**	**185**	**149**	**114**	**36**		

p. 44 (3 digits divided by 2 digits)

1	2	3	4	5	6	7	8	9	10
11	**5**	**18**	**17**	**13**	**7**	**15**	**14**	**23**	**13**
11	12	13	14	15	16	17	18	19	20
14	**15**	**4**	**13**	**23**	**7**	**9**	**12**	**25**	**14**
21	22	23	24	25	26	27	28		
34	**10**	**13**	**7**	**13**	**13**	**17**	**26**		

p. 51 (decimal division)

1	2	3	4	5	6	7	8	9	10
30	**0.9**	**13**	**9**	**15**	**33**	**0.6**	**15**	**9**	**1.5**
11	12	13	14	15	16	17	18	19	20
1.7	**24**	**0.27**	**13**	**22**	**12**	**7.9**	**7.5**	**25**	**28.4**
21	22	23	24	25	26	27	28		
24	**34**	**6.2**	**1.3**	**8**	**6**	**13**	**5.3**		

Love you

Made in United States
North Haven, CT
05 November 2026

21772048R00035